Diseases Of Field Crops And Their Management

The book entitled "Diseases of field crops and their management" provides most recent information about major diseases of cultivation field crops, their symptoms, pathogen characters, epidemiology, and management. In order to make the book all in one, the importance of major diseases has also been dealt with in brief.

Dr. S. Parthasarathy is Assistant Professor (Plant Pathology), College of Agricultural Technology, Theni. He completed his Ph.D. from Tamil Nadu Agricultural University, Coimbatore. He is a recipient of BIRAC UIC Innovation Fellow in DBT during 2015 and was also received several medals and awards in International and National conferences.

Dr. G. Thiribhuvanamala is Associate Professor in Department of Plant Pathology, Tamil Nadu Agricultural University, Coimbatore. She completed her Ph.D. from Tamil Nadu Agricultural University, Coimbatore.

Dr. K. Prabakar, Director, Centre for Plant Protection Studies, Tamil Nadu Agricultural University, Coimbatore, is very sincere and hardworking teacher. He has taught many undergraduate, post-graduate and doctoral courses and guided many PG and Ph.D students over 26 years with great dedication in a dynamic manner.

Diseases Of Field Crops And Their Management

Dr. S. Parthasarathy
Assistant Professor (Plant Pathology)
College of Agricultural Technology, Theni

Dr. G. Thiribhuvanamala
Associate Professor (Plant Pathology)
Tamil Nadu Agricultural University, Coimbatore

Dr. K. Prabakar
Director
Centre for Plant Protection Studies
Tamil Nadu Agricultural University, Coimbatore

NARENDRA PUBLISHING HOUSE
DELHI (INDIA)

First published 2021
by CRC Press
2 Park Square, Milton Park, Abingdon, Oxon, OX14 4RN
and by CRC Press
6000 Broken Sound Parkway NW, Suite 300, Boca Raton, FL 33487-2742

© 2021 Narendra Publishing House

CRC Press is an imprint of Informa UK Limited

The right of S. Parthasarathy, G. Thiribhuvanamala and K. Prabakar to be identified as authors of this work has been asserted by them in accordance with sections 77 and 78 of the Copyright, Designs and Patents Act 1988.

Reasonable efforts have been made to publish reliable data and information, but the author and publisher cannot assume responsibility for the validity of all materials or the consequences of their use. The authors and publishers have attempted to trace the copyright holders of all material reproduced in this publication and apologize to copyright holders if permission to publish in this form has not been obtained. If any copyright material has not been acknowledged please write and let us know so we may rectify in any future reprint.

All rights reserved. No part of this book may be reprinted or reproduced or utilised in any form or by any electronic, mechanical, or other means, now known or hereafter invented, including photocopying and recording, or in any information storage or retrieval system, without permission in writing from the publishers.

For permission to photocopy or use material electronically from this work, access www.copyright.com or contact the Copyright Clearance Center, Inc. (CCC), 222 Rosewood Drive, Danvers, MA 01923, 978-750-8400. For works that are not available on CCC please contact mpkbookspermissions@tandf.co.uk

Trademark notice: Product or corporate names may be trademarks or registered trademarks, and are used only for identification and explanation without intent to infringe.

Print edition not for sale in South Asia (India, Sri Lanka, Nepal, Bangladesh, Pakistan or Bhutan).

British Library Cataloguing-in-Publication Data
A catalogue record for this book is available from the British Library

Library of Congress Cataloging-in-Publication Data
A catalog record has been requested

ISBN: 978-0-367-54041-8 (hbk)
ISBN: 978-1-003-08419-8 (ebk)

NARENDRA PUBLISHING HOUSE
DELHI (INDIA)

Content

	Preface	*ix*
1	Diseases of Rice	1
2	Diseases of Wheat	39
3	Diseases of Barley	65
4	Diseases of Oat	72
5	Diseases of Maize	75
6	Diseases of Sorghum	102
7	Diseases of Pearl millet	120
8	Diseases of Finger millet	127
9	Diseases of Foxtail millet	139
10	Diseases of Indian Barnyard millet	145
11	Diseases of Kodo millet	149
12	Diseases of Proso millet	154
13	Diseases of Little millet	158
14	Diseases of Job's tear	161
15	Diseases of Redgram	162
16	Diseases of Black gram and Green gram	170
17	Diseases of Bengal gram	180
18	Diseases of Horsegram	186
19	Diseases of Soybean	194

20	Diseases of Moth bean	201
21	Diseases of Kidney bean / French bean	207
22	Diseases of Cowpea	210
23	Diseases of Lablab	220
24	Diseases of Indian pea	224
25	Diseases of Lentil	227
26	Diseases of Groundnut	232
27	Diseases of Sesame	247
28	Diseases of Sunflower	256
29	Diseases of Safflower	264
30	Diseases of Castor	269
31	Diseases of Linseed	277
32	Diseases of Mustard	285
33	Diseases of Niger	295
34	Diseases of Cotton	302
35	Diseases of Jute	318
36	Diseases of Silk-cotton	325
37	Diseases of Sunn hemp	327
38	Diseases of Mesta	333
39	Diseases of Agave	337
40	Diseases of Ramie	339
41	Diseases of Sugarcane	343
42	Diseases of Sugarbeet	360
43	Diseases of Tobacco	366

44	Diseases of Jatropha	379
45	Diseases of Mulberry	382
46	Fungal and Bacterial spoilage of grains and storage pathogens	392
	References	*401*

Preface

The book in hand entitled "Diseases of Field Crops and their Management" gives most recent information about major diseases of cultivation field crops, their symptoms, pathogen characters, epidemiology, and management. In order to make the book all in one, the importance of major diseases has also been dealt with in brief.

While preparing the book a number of books, research papers, manuals, IPM modules and online contents on the diseases of field crops are studied and drawn information from those and presented in this book. We wish to express my sincere thanks to my teachers, authors of various book and scientific works of literatures which is the foundation of this book.

The purpose of writing this book was mainly to present before plant pathologists a comprehensive text book of different type of diseases occurring on the plants in the nurseries, fields, controlled conditions, gardens, orchards, plantations etc. We request all the scientists, teachers, experts and students who read this book to comment and intimate me the rectifications and improvement for further validation. Our thanks are due to all those who inspired us in taking up this project.

We hope the book in the present form will fulfill the desired needs of the scientists, teachers and students alike.

Authors

Chapter - 1

Diseases of Rice - *Oryza sativa* L.

S. No.	Disease	Pathogen
1.	Blast	*Magnaporthe oryzae*
2.	Brown spot	*Cochliobolus miyabeanus*
3.	Sheath rot	*Sarocladium oryzae*
4.	Sheath blight	*Thanatephorus cucumeris*
5.	Stem rot	*Magnaporthe salvinii*
6.	False smut	*Ustilaginoidea virens*
7.	Udbatta	*Balansia oryzae-sativa*
8.	Stackburn	*Trichoconiella padwickii*
9.	Narrow brown leaf spot	*Cercospora janseana*
10.	Foot rot	*Gibberella fujikuroi*
11.	Grain discolouration	Fungal complex
12.	Bunt	*Tilletia barclayana*
13.	Leaf smut	*Entyloma oryzae*
14.	Leaf scald	*Microdochium oryzae*
15.	Bacterial blight	*Xanthomonas oryzae* pv. *oryzae*
16.	Bacterial leaf streak	*Xanthomonas oryzae* pv. *oryzicola*

17.	Rice tungro disease	*Rice tungro virus*
18.	Rice grassy stunt disease	*Rice grassy stunt virus*
19.	Rice ragged stunt disease	*Rice ragged stunt virus*
20.	Rice dwarf disease	*Rice dwarf virus*
21.	White leaf disease	*Rice hoja blanca virus*
22.	Rice yellow dwarf disease	*Candidatus* Phytoplasma oryzae
23.	Ufra disease	*Ditylenchus angustus*
24.	Khaira disease	Zinc deficiency

1.   Blast

Other names: Grey leaf spot, Richman disease, Rice fever, Brusone

Significant History

>> The rice blast disease was first reported in china by Soong ying-shin in 1637 in his book utilization of natural resources and it was first reported by Tsuchiya in Japan in 1704 from Italy in 1828 and from USA in 1876.

>> The causal organism, *P. oryzae*, was named by Cavara in Italy (Cavara, 1891) and subsequently in Japan (Shirai, 1896).

>> In India, blast was first recorded during 1913 (Mc Rae, 1922), the disease gained much significance when devastating epidemics observed in Tanjavur Delta region of Tamil Nadu state during 1919 (Padmanabhan, 1965).

>> Later it was recorded in Maharastra in 1923 and subsequently in all rice growing states of the country. In severe cases grain losses of 70 to 80% were reported.

Occurrence

In Kharif season, the disease is prevalent throughout the rice-growing areas in India especially in Himachal Pradesh, Uttarakhand, Jharkhand, Madhya Pradesh, Chattisgarh, Assam, Tripura, West Bengal, Odisha, Maharashtra, Andhra Pradesh, Kerala, Karnataka, Tamil Nadu and Telangana.

In Rabi season, the disease is prevalent in Southern States like Andhra Pradesh, Tamil Nadu and Karnataka. The disease is also common on boro rice in the states of Assam, Tripura, Eastern Uttar Pradesh, Odisha and West Bengal.

Diseases of Rice

Symptoms

» The fungus attacks the crop at all stages of its growth.

» Symptoms appear on leaves, nodes, rachis, and glumes.

» The recent reports shows even roots can become infected.

» All stages of the crop growth (seedling to maturity) were infected.

» Depending on the site of symptom rice blast is referred as leaf blast, collar blast, node blast, and neck or panicle blast and grain infection. Amongst which, neck blast is the most destructive phase of the disease.

Types of blast – based on stage of infection

a. **Seedling blast (nursery)** – Predominant in tropical conditions

b. **Tillering blast (main field)** – Predominant in temperate conditions

Types of blast – based on part of infection

a. **Leaf blast:** The symptoms on the leaves may vary according to the stage of the crop, resistance level and environmental conditions. The lesions may appear as small water soaked grey-green spots with a darker green border and they expand rapidly under moist weather to form the characteristic diamond or spindle shaped spots with grey centre and dark brown margin on the leaf blades. The spots coalesce as the disease progresses and large areas of the leaves dry up and wither. Spots also appear on sheath rarely. Severely infected nursery and field appear as burnt.

b. **Nodal blast:** Besides infecting the leaves, during heading stage the pathogen also invades on nodes of the culms. Almost black color lesions appear on nodes and the tissue appears shrinkage. The affected nodes may break up and all the plant parts above the infected nodes may die.

c. **Collar blast:** Collar blast is seen at the base of the flag leaf. Brown necrotic lesion occurs at the junction of the leaf and the stem sheath. Collar blast will lead to kill the entire leaf.

d. **Neck blast or Rotten neck or Panicle blast:** During flower emergence, lesions can be found on the panicle branches, spikes, and spikelets. Necks are turned to brownish-black. In early neck infection, chaffiness occurs while in late infection, partial grain filling occurs and may not attain normal size. Small brown to black spots may also be observed on glumes of the heavily infected panicles.

e. **Grain infection**: Seeds are fails to produce when pedicels become infected, a condition called blanking. The fungus can infect seeds by infecting the florets as they mature into seeds, and it is believed that this is the main way seed infection develops. The infected individual spikelet shows greyish discoloration on lemma and palea.

Fungus: Teleomorph: *Magnaporthe oryzae* (B.C. Couch)
Anamorph: *Pyricularia oryzae* (Cavara) (Early: *Pyricularia grisea* (Cooke) Saccardo)
(Early: *Trichothecium griseum* Cooke; *Dactylaria oryzae* Cav.)

Fungal characters

Anamorph:

- » The mycelium is hyaline to olivaceous-brown and septate.

- » Conidiophores emerge from the stomata or epidermal cells, singly or in clusters.

- » Conidia are produced in clusters on apex of the long septate, olivaceous sympodial conidiophores. Conidia are attached at the broader base by a hilum. Conidia are haploid, pyriform (*Pyricularia*) to ellipsoid, hyaline to pale olive green. The conidia are usually three-celled (two septate), 14-40 μm in length and 6-15 μm in width. Conidia are produced after several hours of high humidity and are easily released or liberated near mid-day, especially under windy conditions.

- » The fungus produces terpenoid phytotoxins called pyricularin.

Teleomorph:

- » The perfect state of the fungal genus is *Magnaporthe* producing perithecia.

- » The asci are unitunicate.

- » The ascospores are hyaline, fusiform (spindle-shaped with tapering ends), four celled by producing three septa and slightly curved.

Diseases of Rice

Disease cycle

The disease spreads primarily through airborne conidia since spores of the fungus present throughout the year. Mycelium and conidia in the infected straw and seeds are major sources of inoculum. Irrigation water may carry the conidia to different fields.The fungus also survives on collateral hosts *viz., Panicum repens, Digitaria marginata, Digitaria sangunalis, Brachiaria mutica, Leersia hexandra, Dinebra retroflexa, Setaria intermeidia* and *Echinochloa crusgalli.*

Spores land on leaves, germinate, penetrate the leaf, and cause a lesion 4 days later; more spores are produced within 6 days. Infections from spores arriving from a distance are termed primary infections. Primary infections generally result in a few widely scattered spots on leaves. Spores arising from the primary infections are capable of causing many more infections. This cycling is called secondary spread. Secondary spread is responsible for the severe epidemics of blast in fields and localized areas.

Favourable conditions

- » Intermittent drizzles, cloudy weather, more of rainy days, longer duration of dew high relative humidity (93-99 %).

- » Low night temperature (nearly 15-20°C or <26°C).

- » Availability of collateral hosts nearby the field.

- » Application of excess nitrogen.

- » Extended leaf wetness period (>10 hrs).

- » Degree of host susceptibility.

Forecast for rice blast can be made on the basis of minimum night temperature range of 20-26°C in association with a high relative humidity of 90% and above lasting for a period of a week or more during any of the three susceptible phases of crop growth, *viz.,* seedling stage, post transplanting tillering stage and neck emergence stage. In Japan, the first leaf blast forecasting model was developed named as BLAST. Later several other models have also been developed namely, PYRICULARIA, PYRIVIEW, PYRNEW, BLASTL, BLASTCAST, BLASTAM, LEAFBLST, EPIBLA, BLASTSIM.2, BLASTMUL, EPIRICE and PBLAST.

Economic threshold level: Mid-tillering to booting: 5-10 % leaf area damage
Flowering to after: 5% leaf area damage

Management

» Remove and destroy the weed hosts in the field bunds and channels.

» Use a mixture of resistant varieties and include susceptible ones in your crop to avoid breaking of host resistance.

» Carry out crop rotation, intercropping, and/or fallow.

» Grow resistant to moderately resistant varieties CR 1009, IR 20, IR 34, IR 64, Co 43, Co 44, Co 47, ADT 20, ADT 36, ADT 39, ADT 40, ASD 18, Bahadur, Ranjit, Luit, Chilarai, Chiana Bora, Jyoti Pradad, Bishnu Prasad, A-67, A-90, A-249, MTU 9992, NLR 9672, NLR 9674, IET 2222, IET 826, IET 2233, Pusa 2-21, Sravani, Snaha, Bani, Mandavijaya, Jaya, Vijaya, Ratna, Pankaj, Jayanti, Pusa 743, Rashi, Prasanna, Saket, Vikas and Tulsi. Avoid cultivation of highly susceptible varieties *viz.,* IR 50 and TKM 6 in disease favourable season.

» Modify plant density so to avoid close spacing since this favors warm and humid conditions ideal for blast. Varieties with high tillering should be planted more separately compared to low tillering varieties.

» Provide a balanced nutrition to the rice crop.

» Silicon fertilizers (e.g., calcium silicate) can be applied to soils that are silicon deficient to reduce blast.

» Seed treatment with captan or carbendazim 50 WP or tricyclazole 75 WP at 2-3 g/ kg or *Pseudomonas fluorescens* (Pf1) at 10 g/ kg.

» Spray the nursery with carbendazim 50 WP 0.2% or tricyclazole 75 WP 0.1% or 300-400 g/ ha.

» Spray the main field with organophosphorous fungicides edifenphos (hinosan) 1 ml/ lit or iprobenphos (kitazin) 48% EC 1 ml/ lit.

» Spray the main field with melanin biosynthesis inhibitors or antipenetrants tricyclazole 0.6 g/ lit. or isoprothiolane 40% EC 750 ml/ ha or kresoxim-methyl 44.3% SC 500 ml/ ha or tebuconazole 25.9% EC 750 ml/ ha or metominostrobin 20 SC at 500 ml/ ha or tebuconazole 25.9% + trifloxystrobin 25% WG 200 g/ ha or azoxystrobin 25 SC at 500 ml/ ha or tricyclazole 22% + hexaconazole 3% SC.

» Spray the main field with carpropamid 27.8% SC at 0.1% (inhibit melanin biosynthesis for blast and anthracnose diseases).

Diseases of Rice

» Spray the main field with antipiriculin A, blastimycin, blasticidin-S WP 20 ppm or kasugamycin WP 20 ppm.

2. Brown spot

Other names: Helminthosporiose, *Helminthosporium* leaf spot, Sesame leaf spot, Fungal blight, Poor man's disease.

Significance

Brown spot is problem mainly during kharif season especially in uplands and hill ecosystem. The disease also assumes a serious proportion in irrigated ecosystem especially in ill-managed plots.

» The pathogen of brown spot of rice was first reported by Breda de Haan in 1900.

» The first report of this disease came from Madurai under Madras Presidency in 1919 (Sundararaman, 1919), during 1942-43 due to the severe damage caused by this disease to paddy famine occurred in Bengal and then it is reported from all rice growing states (Padmanabhan, 1973).

» The havoc of brown spot disease leads to the Great Bengal famine in India.

Symptoms

a. **Leaf spot**: The fungus attacks the crop from seedling to milky stage in main field. Symptoms appear as minute spots on the coleoptile, leaf blade, leaf sheath, and glume, being most prominent on the leaf blade and glumes. The spots become cylindrical or oval, dark brown sesame seed shaped (2-5×0.5-2 mm) with yellow halo later becoming circular. Several spots coalesce and the leaf dries up. The seedlings die and affected nurseries can be often recognized from a distance by scorched appearance.

b. **Grain discoloration**: In later stage, the grain becomes shrivelled. Panicle turns dark brown or black. The grains are brown or black spots also appear on glumes leading to discoloration. The affected grains are covered with olive-velvet fungal growth. It causes failure of seed germination, seedling mortality and reduces the grain quality, germination percentage and weight.

Fungus: Teleomorph: *Cochliobolus miyabeanus* (Ito and Kuribayashi)
Anamorph: *Bipolaris oryzae* (Breda de Haan)

Syn: *Helminthosporium oryzae* (Breda de Haan, 1900), *Drechslera oryzae* (Dreshler).

Fungal characters

Bipolaris oryzae produces brown, septate mycelium. Conidiophores arise singly or in small groups. They are geniculate, brown in colour. Conidia are usually curved with a bulged center and tapered ends. They are pale to golden brown in colour and are 6-14 septate. The perfect stage of the fungus is *C. miyabeanus.*

It produces perithecia with asci containing 6-15 septate, filamentous or long cylindrical, hyaline to pale olive green ascospores. The fungus produces terpenoid phytotoxins called Ophiobolin A (or Cochliobolin A), Ophiobolin B (or Cochliobolin B) and Ophiobolin I. Ophiobolin A is most toxic. These breakdowns the protein fragments of cell wall resulting in partial disruption of integrity of cell.

Disease cycle

Infected seeds and stubbles are the most common source of primary infection. The conidia present on infected grain and mycelium in the infected tissue are viable for 2 to 3 years. Airborne conidia infect the plants both in nursery and in main field.

The fungus also survives on collateral hosts like *Leersia hexandra* and *Echinochlona colonum.* The brown spot fungus is normally present in areas with a long history of rice cultivation. Airborne spores that are capable of causing infection are produced in infested debris and older lesions.

Favourable conditions

» Temperature of 25-30°C with relative humidity above 80% is highly favourable.

» Excess of nitrogen aggravates the disease severity.

» The fungus survives in *Setaria italica, Panicum repens, Leersia hexandra* and *Echinochloa colona, E. crusgalli* and *Cyanodan dactylon.*

» Occurs mostly in nitrogen and potassium deficient and poor soils.

Economic threshold level: 2.5% affected tillers

Diseases of Rice

Management

» Field sanitation, removal of collateral hosts and infected debris from the field.

» Grow resistant varieties *viz.,* Amruth, Bala, Bhavani, Padma, Rasi, Tellahamsa, Kakatiya, Co 44, IR 24, IR 36, Gajraj, Jaya, Ratna, Rambhog, Sona, Krishna, Sabarmathi and Jagannath.

» Use of slow release nitrogenous fertilizers and timely application of potassic fertilizer as top dressing is advisable.

» Seeds can also be treated with hot water at 55°C for min.

» Seed treatment with captan at 4 g/ kg.

» Spray the nursery with edifenphos 40 ml or mancozeb 80 g for 20 cent.

» Spray the main field with *Pseudomonas fluorescens* 5-10 g/ lit.

» Spray the main field with aureofungin 46.15% SP 0.005% or tricyclazole 18% + mancozeb 62% WP 1 kg/ ha or edifenphos 500 ml/ ha or mancozeb 2 kg/ ha or propineb 70% WP 2 kg/ ha or metominostrobin 500 ml/ ha mostly at tillering and flowering stage. If needed repeat after 15 days.

3. Sheath rot

The disease was first identified by Sawada from Taiwan in 1922 and later it became a common problem in Japan. Sheath rot is present in most of the rice growing countries worldwide, particularly in rainfed rice ecosystems and is more prevalent during wet than dry seasons. Sheath rot infection results in chaffy, discoloured grains and affects the viability and nutritional value of seeds.

Symptoms

» Initial symptoms are noticed only on the upper most leaf sheath enclosing young panicles.

» The flag leaf sheath showed oblong or irregular greyish brown spots.

» They enlarge and develop grey centre and brown margins covering major portions of the leaf sheath.

» The young panicles remain within the sheath or emerge partially.

> The panicles rot and abundant whitish powdery fungal growth is seen inside the leaf sheath.

Fungus: Anamorph: *Sarocladium oryzae* (Sawada) Gems et Hawksworth
Syn: *Sarocladium attenuatum* (W. Gams and D. Hawksw);
Acrocylindrium oryzae (Sawada)

Fungal characters

The genus *Sarocladium* was established in 1975. It is a saltatory morphological apomorph. No teleomorphs are known to be associated with this group. The fungus produces whitish, sparsely branched, septate mycelium. Mycelial colonies of *S. oryzae* on the host are pale pinkish or whitish and sporulation is abundant in the center. *S. oryzae* produces verticilliate condiophore with 1 or 2 phialides which contain fusiform, often curved, hyaline, single cell conidia of 3.5-9×1-2.5 mm. Conidia are hyaline, smooth, single celled and cylindrical in shape.

> *Sarocladium oryzae* produces toxins *viz.*, Helvolic acid which induce chlorosis and Cerulenin induce necrosis and growth inhibition

Disease cycle

The disease spreads mainly through air-borne conidia and also seed-borne.

> Primary source of inoculum is by means of infected plant debris.

> Secondary spread is by means of air borne conidia produced on the leaf sheath.

Tersonemid mite, *Brevennia rehi* (Mealybug) and *Leptocorisa acuta* (Ear head bug) facilitates the spread of the disease in India.

Worldwide disease etiology of rice sheath rot and mainly deals with the three most reported rice sheath rot pathogens: *Sarocladium oryzae*, the *Fusarium fujikuroi* complex, and *Pseudomonas fuscovaginae*.

Favourable conditions

> Closer planting, nitrogen-responsive, high yielding semi-dwarf as well as tall rice varieties.

> Photoperiod-sensitive tall varieties were more resistant than photoperiod-insensitive varieties.

Diseases of Rice

» High humidity and temperature around 25-30°C.

» Injuries made by leaf folder, brown plant hopper and mites increases the infection.

Economic threshold level: Flowering and after 2-5% affected tillers.

Management

» Grow resistant varieties like Tetop, IR 50, Ratna, Sakti, Polman-579, MR 1553, IET 826 and moderately resistance varieties like ADT 39, Co 47 and TKM 13.

» Apply potash at tillering stage.

» Soil application of gypsum 500 kg/ ha in two splits.

» Seed treatment with carbendazim 0.2%, edifenphos 0.1%, or mancozeb 0.2% and foliar spray at booting stage.

» Foliar spray with calcium sulfate and zinc sulfate.

» Spray with Neem Seed Kernal Extract (NSKE) 5% or Neem oil 3% or Neem soap, *Ipomoea* or *Prosopis* leaf powder extract 25 kg/ ha. first spray at boot leaf stage and second 15 days later.

» Spray the main field with *Pseudomonas fluorescens* 5-10 g/ lit.

» Spray carbendazim 500 g or edifenphos 1 lit or mancozeb 2 kg/ ha or metominostrobin at 500 ml/ ha or carbendazim 27% + mancozeb 68% WP 20-2.5 g/ lit. at boot leaf stage and 15 days later.

4.   Sheath blight

Other names: Banded leaf blight, Earhead choaking, Snake skin disease.

Significance

Sheath blight of paddy is one of the most widely spreading diseases of paddy. This disease was first time reported from Japan in 1910 by Miyaki. In India, Paracer and Chahal reported this disease from Gurdaspur (Punjab) only in 1963. They described the disease in detail and the causal fungus was identified as *Rhizoctonia solani*. Singh and Pavgi (1969) described the development of perfect stage of the fungus on rice plants for the first time in India and the pathogen was identified as *Corticium sasakii* (Shirai) Matsumoto. Talbot (1970) considered *Thanatephorus cucumeris* (Frank) Donk to be the perfect state.

This disease is problematic in areas where irrigation facilities are abundant. Intensive and changed cultivar practices have intensified the severity of the disease. It is a potentially devastating disease of rice in all temperate and tropical rice production regions, especially in irrigated production systems. This disease severely occurs in Uttar Pradesh, Kerala and parts of Tamil Nadu.

Symptoms

» Initial symptoms are noticed on leaf sheaths of lower most leaves near the water line when plants are in the late tillering or early internode elongation stage (approximately 10-15 days after flooding) varies from place to place.

» On the leaf sheath oval or elliptical or irregular greenish grey spots (1-3 cm long) are formed about ¼ inch wide and ½ to 1 ¼ inch long. As the spots enlarge, the centre becomes greyish white with an irregular blackish brown or purple brown border.

» Disease development progresses very rapidly in the during periods of frequent rainfall and overcast skies.

» Lesions on the upper parts of plants extend rapidly coalescing with each other to cover entire tillers from the water line to the flag leaf.

» The presence of several large lesions on a leaf sheath usually causes death of the whole leaf, and in severe cases all the leaves of a plant may be blighted.

» The infection extends to the inner sheaths resulting in death of the entire plant.

» Sclerotia, initially white turning dark brown at maturity, are produced superficially on or near the lesions.

» Sclerotia are loosely attached and easily dislodge from the plant.

Fungus: Teleomorph: *Thanatephorus cucumeris* (A.B. Frank)
Anamorph: *Rhizoctonia solani* AG-1(J.G. Kühn)

Fungal characters

The fungus produces septate mycelium which are hyaline when young, yellowish brown when old. It produces lateral hyphae arise at right angles from the main hyphae, and possess minute constriction at the point of origin or attachment. During sexual reproduction, it produces barrel shaped basidia.It produces large number of spherical brown sclerotia.

Diseases of Rice 13

Disease cycle

The mycelium enters into host plant through stomata or the cuticle. The pathogen can survive as sclerotia or mycelium in dry soil for about 20 months but for 5-8 months in moist soil. Sclerotia spread through irrigation water. The fungus has a wide host range.

» This is the primary source of inoculum from the soil.

» The sclerotia float to the surface of the water during soil puddling, levelling, weeding operations and infect the plants on contact.

» The floating water acts as secondary source of infection through sclerotia and mycelium.

» The collateral hosts for this pathogen is *Sateria glauca, Cyanodon dactylon, Echinocloa colona, E. crusgalli, Panicum ripens, Cyperus rotundus, C. irrea, Zea mays* and *Eleusine coracana.*

Favourable conditions

» High relative humidity (85-100%) and high temperature (28-32°C).

» Closer planting.

» Heavy doses of nitrogenous fertilizers.

» High seeding rate or close plant spacing, dense canopy, disease in the soil, sclerotia or infection bodies floating on the water, and growing of high yielding improved varieties also favor disease development.

» Sheath blight is a fungal disease, more common in rainy season especially early heading and grain filling stages especially in the lower part of the panicle.

Management

» Grow resistant varieties like Saket-1, Krishna, Mansarovar, Swarau Dhan, Pankaj, Swarnadhan, Vikramarya, Radha, Mandya, Shiva and Vijaya.

» Avoid sowing of hybrids and water logging condition.

» High potassium induces resistance to the disease.

» Apply organic amendments *viz.,* neem cake at 150 kg/ ha or FYM 12.5 tonnes/ ha or *Diancha.*

14 Diseases of Field Crops and their Management

» Avoid flow of irrigation water from infected fields to healthy fields.

» Foliar spray with Neem oil at 3% (15 lit/ha) starting from disease appearance.

» Deep ploughing in summer and burning of stubbles.

» Soil application of *P. fluorescens* at of 2.5 kg/ ha after 30 days of transplanting (product should be mixed with 50 kg of FYM/ ha applied).

» Soil application of thiram or brassical 25 kg/ ha.

» Foliar spray *P. fluorescens* at 0.2% at boot leaf stage and 10 days later.

» Spray carbendazim 500 g/ ha or carbendazim 25 + flusilazole 12.5% SE 1 kg/ ha or difenoconazole 25% EC 500 ml/ lit or flusilazole 40% EC 300 ml/ ha or hexaconazole 5% EC 1000 ml/ ha or iprodione 50% WP 2.25 kg/ ha or kitazin 48% EC 200 ml/ lit or pencycuron 22.9% SC 750 ml/ ha or thifluzamide 24% SC 375 g/ ha or tebuconazole 50% + trifloxystrobin 25% WG 200 gm/ ha.

» Spray with validamycin 3% L 30-60 ppm of 2 kg/ ha or polyoxin AL 10% AL 20 ppm of 2 kg/ ha.

5.   Stem rot

Stem rot leads to formation of lesions and production of chalky grains and unfilled panicles. This disease was first reported by Cattaneo 1976 from Italy. In India occurance was first noticed in 1911 and described by Butler 1918.

Symptoms

» Small black lesions are formed on the outer leaf sheath and they enlarge and reach the inner leaf sheath also.

» The affected tissues rot and abundant small black sclerotia are seen in the rotting tissues.

» Typical symptoms are apparent in this disease affected area as the crop approaches maturity the plants lose, turn yellow and finally die.

» The culm collapses and plants lodge.

Fungus: Teleomorph: *Magnaporthe salvinii* (Catt.) R.A. Krause & R.K. Webster
   Syn: *Nakataea sigmoidae* (Cavara) K. Hara
   Anamorph: *Sclerotium oryzae* (Cattaneo)
   Syn: *Leptosphaeria salvinii* (Cattaneo)

Diseases of Rice 15

Fungal characters

The mycelium is white to greyish color, profusely branched. The conidiophore is dark and septate. Conidia are fusiform, sigmoid and three septate. The sclerotia are spherical black and shiny sclerotia, visible to naked eyes as black masses. The perithecia (sexual) are globose with beak, produces clavate asci containing eight ascospores.

Disease cycle

The sclerotia survive in stubbles and straw those are carried through irrigation water. The fungus over winters and survives for long periods as sclerotia in the upper layers (2-3 inches) of the soil profile. The half-life of sclerotia in the field is about 2 years and viablity of sclerotia remains up to 6 years. The sclerotia are buoyant and float to the surface of floodwater where they contact, germinate, and infect rice tillers near the water line.

Favourable conditions

»   The disease is favored by low soil potassium and high nitrogen levels.

»   Stem rot is serious in fields that have been in continuous rice for several years.

»   Infestation of leaf hoppers and stem borer.

Management

»   Deep ploughing in summer and burning stubbles to eliminate sclerotia.

»   Grow resistant varieties like Jaya, Vijaya, IR 22, IR 36, IR 64, IET 1136, WC 679, NC 1281 and WC 1281.

»   Split application of nitrogenous, potassic fertilizers lime to increase soil pH at basal dose, tillering stage and panicle initiation stage.

»   Avoid flow of irrigation water from infected to healthy fields.

»   Draining irrigation water and letting soil to dry.

»   Spot drenching with cerason 1 g/ lit. or captan 2.5 g/ lit. or *Trichoderma viride* talc 4 gm / plant under semi-dry condition.

»   Spray with thiaphanate methyl 0.2% and validamycin A 3% L 2.6 g/ lit. at the mid-tillering stage.

6. False smut

Other names: Green smut, Orange smut, Laxmi disease, Bumper disease

False smut causes chalkiness of grains which leads to reduction in grain weight. It also reduces seed germination. The disease was first reported from Tirunelveli district of Tamil Nadu State of India (Cooke, 1878). Historically being an uncommon and minor disease by occurring sporadically in certain regions.

Symptoms

» Plants infected with false smut have individual rice grain transformed into a velvety mass of spore balls (pseudomorph) due to penetration and spread of pathogen in the flower ovary.

» The false smut ball is covered with powdery chlamydospores, and the colour changes to yellowish, yellowish orange, green, olive green and, finally, to greenish black.

» In most cases, not all spikelets of a panicle are affected, but spikelets neighboring smut balls are often unfilled or sterile. Only a few spikelets in a panicle are affected.

» Indeed, in some areas this disease is more prevalent in seasons favourable for good growth and high yields, and the farmers consider its incidence as an omen of good harvest.

Fungus: Anamorph: *Ustilaginoidea virens* (Cooke, 1975) Tahahashi
(Syn: *Claviceps oryzae – sativa* Hashioka)
Teleomorph: *Villosiclava virens* (Y. Sakurai ex Nakata)

Fungal characters

It possesses a teleomorphic state producing sexual ascospores and an anamorphic state generating asexual chlamydospores. Conidia are round to elliptical. Chlamydospores are ornamented with prominent irregularly curved spines. The sclerotia are black, horseshoe-shaped and irregular oblong or flat. *U. virens* produces a toxin, known as Ustiloxin.

Favorable conditions

» Rainfall and cloudy weather during flowering and maturity.

» Low temperature (20°C).

Diseases of Rice 17

» High relative humidity (>92%).

» Moderate rainfall with intermittent clear and drizzling weather during flowering.

» More prevalent in seasons favorable for good growth and high yields.

» The pathogen also survives through alternate host *viz.*, barnyard grass (*Echinochloa crusgalli*), *Imperata cylindrica*, and common rice weed *Digitaria marginata*.

Disease cycle

» At the start, the balls consist of white hyphae, which then form a thick, yellow, loose outer layer of chlamydospores (spherical and warty) in summer and early autumn, and an olive to black, hard outer layer in late autumn. Sclerotia often form on the colony surfaces, especially in later autumn, with lower temperatures and high temperature differences between day and night.

» The disease affects the early flowering stage of the rice crop when the ovary is destroyed. The second stage of infection occurs when the spikelet nearly reaches maturity. Grasses and wild rice species are alternate hosts. The fungus overwinters in soil by means of sclerotia and chlamydospores. Sclerotia produce ascospores, which are the primary source of infection to rice plants, whereas secondary infection comes from airborne chlamydospores.

» The pathogen is internally seed borne.

Management

» Deep ploughing in summer and burning of stubbles.

» Grow resistant varieties like Saket 4, IR 26, IR 8, Surya, IR 20, Masuri, Pankaj, Sakti, Sabarmati and Vijaya.

» Use disease free seeds for sowing.

» Treat seeds at 52°C for 10 min.

» Use moderate rates of nitrogen.

» Removal and destruction of diseased panicles in field.

» Seed treatment with carbendazim 2.0g/ kg.

» Spray hexaconazole at 1 ml/ lit or chlorothalonil 2 g/ lit or copper hydroxide 77% WP

2 kg/ ha or copper oxychloride 0.25%, tebuconazole 0.1%, propiconazole 25% EC, difenoconazole 0.1% and validamycin A 2.6 g/ lit at tillering and preflowering stages.

7. Udbatta

Other names: False ergot, Agarbatti disease and Spike rot

The disease was first detected in Nongpoh, Meghalaya. This disease is endemic and of minor importance in certain areas in India. Infection of the plant is systemic and results in the loss of most or total yield.

Symptoms

- » The systemic infection, results in the emergence of an erect, greyish-white, cylindrical axis from the leaf sheath instead of a normal inflorescence.
- » The entire ear head is converted into a straight compact cylindrical black spike like structure since the infected panicle is matted together by the fungal mycelium.
- » The spikelets are cemented to the central rachis and the size is remarkably reduced.
- » The entire spike is covered by greyish stroma with convex pycnidia immersed inside.
- » No grains were formed.

Fungus: Teleomorph: *Balansia oryzae-sativa* (Hashioka 1971)
Anamorph: *Ephelis oryzae* (Sydow, 1914) - Spermatial stage

Fungal characters

Stromata on mummified spikelets are hemispherical, capitate, with a black, coarsely papillate surface, yellowish-brown to white interior, arising from a mummified spikelet upon which the conidiamata developed. Ascomata with perithecia embedded in periphery stromatal head, perithecia rounded, ovoid to pyriform. Asci are cylindrical, with a rounded, thickened apex, an attenuated base, eight-spored. Ascospores are filiform, non-septate, straight or curved.

Acervuli on surface of mummified inflorescence emerging from leaf sheath; when wet, appearing gelatinous, cupulate or convex fructification, bearing a palisade of conidiophores. Conidiophores are terminating in narrow conidiogenous cells that proliferate percurrently to form a mass of filiform to acicular, hyaline conidia.

Diseases of Rice

Disease cycle

The primary source of inoculum is considered to be infected seeds.

Favourable conditions

An average soil temperature of 28°C and abundant soil moisture in the nursery beds during the first week after sowing, followed by average soil temperatures of 28-38°C and adequate soil moisture in the subsequent period of growth up to flowering, is conducive to the development of the disease. The collateral hosts *viz.*, Grasses, *Isachne, elegans, Cynadon dactylon, Pennisetum* sp. and *Eragrostis tenuifolia.*

Management

- » Grow tolerant variety J 1 and CB 11.
- » Seed treatment with carbendazim 0.1% and spraying of same fungicide at the panicle initiation stage.
- » Hot water seed treatment at 50-54°C for 10 min effectively controls the disease.
- » Removal of collateral hosts.
- » Spray aureofungin 0.005%, iprodione 6 g/ lit. and mancozeb 75% WP 2 kg/ ha.

8. Stackburn

Other names: *Alternaria* seedling blight and leaf spot

Alternaria padwickii is an asexually reproducing fungus that infects seeds of rice. It is one of several fungi responsible for seed discoloration, seed rot and seedling blight, but has also been detected as a sheath-rotting pathogen. Tisdale (1922) and Tullis (1936) reported on a rice-infecting and sclerotium-producing *Alternaria* in the USA as *Trichoconis caudata* (Appel & Strunk) Clem.

Symptoms

- » This seed-borne fungus causes pre- and post-emergence seed rot.
- » After emergence, small dark brown lesions may occur on the roots, the coleoptile or the early leaves.

20 Diseases of Field Crops and their Management

» The parts of the seedling above lesions are blighted or the whole seedling dies.

» Leaves and ripening grains are affected. On leaves circular to oval spots with dark brown margins are formed.

» The center of the spot turns light brown or white with numerous minute dots.

» In the "stackburn" phase of the disease, spots on glumes are pale brown to white or faintly pink or reddish-brown, usually with a darker border.

» Infected grain is dark colored, chalky, brittle, and/or shriveled, with reduced viability.

» The small black sclerotia appear in the center of lesions on all infected parts.

Fungus: Anamorph: *Trichoconiella padwickii* (Ganguly) B.L. Jain
Syn: *Trichoconis padwickii* (Ganguly, 1947) [Early name]
Syn: *Alternaria padwickii* (Ganguly) M.B. Ellis 1971

Fungal characters

Conidiophores are solitary, unbranched, and smooth. Conidia single, fusiform to obclavate, with filamentous true beak, hyaline to straw or golden brown color, with 3-5 transverse septa, often constricted at septa, smooth or minutely echinulate. Sclerotia are spherical, black, and multicellular, walls reticulate.

Spread

Feeding by rice bugs such as *Leptocorisa oratorius*, *Leptocorisa acuta* and rice stink bug, *Oebalus pugnax* was considered to enhance infection.

Management

» Seed treatment with captan or mancozeb or carbendazim 2-3 g/ kg or *Pseudomonas fluorescens* 5.0 g/ kg.

» Burn the stubbles and straw in the field.

» Hot water treatment at 54°C for 15 minutes is also effective.

» Spray with chlorothalonil 0.2%, mancozeb 0.2%, carboxin 0.3%, fenapanil 0.3%, polyoxin 20 ppm, edifenphos 0.1% and iprodione 0.2%.

Diseases of Rice

9. Narrow brown leaf spot

Raciborski in 1900 first reported this pathogen and named it *Napicladium janseanum.*

Symptoms

» The fungus produces short, linear brown spots mostly on leaves and also on sheaths, pedicels and glumes.

» They are usually 2-10 mm long and 1-1.5 mm wide. The spots appear in large numbers during later stages of crop growth.

» It leads to premature death of leaves and leaf sheaths, premature ripening of grains, and in severe cases, lodging of plants.

» The disease also causes discoloration on the leaf sheath, referred to as "net blotch" because of the netlike pattern of brown and light brown to yellow areas.

» Grain infection can cause a purplish-brown discoloration of the seeds or grain.

Fungus: ***Anamorph:*** *Cercospora janseana* (Racib)
Syn: *Cercospora oryzae* (I. Miyake)
Teleomorph: *Sphaerulina oryzina* (Hara)

Conidiophores are produced in groups and brown in colour. Conidia are hyaline or sub hyaline, cylindrical and 3-5 septate. *C. janseana* produces red to maroon pigmented photo activated perylene quinone toxin - cercosporin.

Disease cycle

The cycle begins when *C. janseana* enters the plant tissues through stomata, establishes beneath the stomata in the parenchyma cells, and spreads longitudinally in intercellular spaces. Conidiophores emerge through the stomata. Studies show that 30 or more days are required to develop symptoms after inoculation. This long latent period may be the probable reason of late appearance of symptoms.

Favourable conditions

» The disease usually occurs in potassium deficient soils and in areas with temperature ranging from 25-28°C at heading stage and most susceptible during panicle initiation onwards to maturity.

22   Diseases of Field Crops and their Management

»   Except rice, the fungus can survive on *Panicum maximum* (guinea grass) *P. repens* (torpedo grass), and *Pennisetum purpureum* (elephant grass).

Management

»   Use balanced nutrients; make sure that adequate potassium is used.

»   Spray with mancozeb 2 kg/ha or propiconazole 25 EC at 0.1% at booting to heading stages.

10.   Foot rot

Other names: Bakanae disease, Foolish seedling disease

The earliest known report of bakanae is from 1828; it was first described scientifically in 1898 by Japanese researcher Shotaro Hori, who showed that the causative agent was fungal. Detailed description of pathogen was made by Kurosawa. In India, it was reported by Thomas 1931.

Symptoms

Nursery stage

»   Sown grains may fail to germinate or the radicle fails to emerge above the soil.

»   Infected seedlings in nursery are lean and lanky, much taller and die after some time.

Main field stage

»   The affected plants have tall lanky tillers with longer internodes and aerial adventitious roots from the nodes above ground level.

»   The root system is fibrous and bushy.

»   The plants are killed before earhead formation or they produce only sterile spikelets. When the culm is split open white mycelial growth can be seen.

»   The diseased plants survive but bear empty panicles.

»   *Gibberella fujikuroi* may also produce perithecia under certain conditions.

»   The afflicted plants, which are visibly etiolated and chlorotic, are at best infertile with empty panicles, producing no edible grains; at worst, they are incapable of supporting their own weight, topple over, and die.

Diseases of Rice

Fungus: Teleomorph: *Gibberella fujikuroi* (Sawada) Ito (Wollenw, 1931)

Anamorph: *Fusarium moniliforme* Sheldon & Nirenberg

Fungus produces both macroconidia and microconidia. Microconidia are hyaline, single celled and oval. Macroconidia are slightly sickle shaped, and two to five celled. The fungus produces the phytotoxin, fusaric acid, which is non-host specific. *Gibberella fujikuroi*, produces a surplus of gibberellic acid. In the plant, it acts as a growth hormone, causing hypertrophy. The fungus is externally seed-borne.

Favourable conditions

» The fungus has a wide range of temperature for optimum growth between 25-35°C.

» Excess use of nitrogen fertilizer predisposes the plant to the attack of pathogen.

Management

» Grow resistant varieties like Ratna, Jaya, Vijaya, Sona, IR 36, Pankaj, Masuri, WC 678, Sakti, MR 1550, Co 18 and ADT 8.

» Clean seeds should be used to minimize the occurrence of the disease.

» Top dressing with postassic fertilizers.

» Seed treatment using fungicides such as thiophanate-methyl 2 g/ kg or carbendazim 2.5 g/ kg or tolyl mercury acetate (Agrosan GN/ Ceresan) 2 g/ kg or thiram 2 g/ kg is effective before planting.

» Steeping seeds in $CuSO_4$ 1% solution or formalin 2% also recommended.

» Seedling dipping with carbendazim 2 g/ lit or ediphenphos 1 ml/ lit or propiconazole 1 ml/ lit or hexaconazole 2 ml/ lit.

11. Grain discolouration

Symptoms

» The grains may be infected by various organisms before or after harvesting causing discoloration, the extent of which varies according to season and locality.

» The infection may be external or internal causing discoloration of the glumes or kernels or both.

» Dark brown or black spots appear on the grains.

» The discoloration may be red, yellow, orange, pink or black, depending upon the organism involved and the degree of infection.

» This disease is responsible for quantitative and qualitative losses of grains.

Fungi: *Drechslera oryzae, D. rostratum, D. tetramera, Curvularia lunata, Trichoconis padwickii, Sarocladium oryzae, Alternaria tenuis, Fusarium moniliforme, Cladosporium herbarum, Epicoccum purpurascens, Cephalosporium* sp., *Phoma* sp. and *Nigrospora* sp.

Disease cycle

The disease spreads mainly through air-borne conidia and the fungus survives as parasite and saprophyte in the infected grains, plant debris and also on other crop debris.

Favourable conditions

» High humidity and cloudy weather during heading stage

Management

» Pre and post-harvest measures should be taken into account for prevention of grain discolouration.

» Seed treatment with captan or mancozeb or carbendazim 2 g/ kg.

» Spray the crop at bootleaf stage with mancozeb 1 kg or Iprobenphos 500 ml or carbendazim 250 g/ ha.

» Store the grains with 13.5-14% moisture content.

12. Bunt

Other names: Kernel Smut, Black smut

Fungus: *Tilletia barclayana* (Bref.) Sacc. & Syd. Syn: *Neovasia horrida* Bref.

Symptoms

Minute black pustules or streaks are formed on the grains which burst open at the time of ripening. The grains may be partially or entirely replaced by the fungal spores. The sorus

Diseases of Rice

pushes the glumes apart exposing the black mass of spores. Only a few flowers are infected in an inflorescence. The fungus survives as chlamydospores for one or more years under normal condition and 3 years in stored grains.

13.   Leaf smut

Fungus: *Entyloma oryzae* (Syd. & P. Syd.). Syn: *Eballistra oryzae* (Syd. & P. Syd.).

Symptoms

Many spots can be found on the same leaf, but the spots remain distinct from each other. Typical symptoms of leaf smut breaks open when wet and releases the black spores. Heavily infected leaves turn yellow, and the leaf tips die and turn gray.

14.   Leaf scald

Fungus: *Microdochium oryzae* (Thum.) Hern.-Restr. & Crous

Symptoms

Zonate lesions of alternating light tan and dark brown starting from leaf tips or edges. Lesions are oblong with light brown halos in mature leaves. Later coalescing of lesions, result in blighting of a large part of the leaf blade. The affected areas dry out giving the leaf a scalded appearance. Infected leaf tips also split near the midrib especially when there is strong wind.

Management of smut and scald

- »   Avoid high use of fertilizer.
- »   Seed treatment with carbendazim, quitozene, chloronil and thiophanate-methyl can be used to to eliminate the disease.
- »   Spray fentin acetate 2 ml/ lit, edifenphos 1 ml/ lit, captafol 2 g/ lit, mancozeb 2 g/ lit, copper oxychloride 2.5 g/ lit and validamycin 30 ppm.

15.   Bacterial blight

Other name: White withering disease

It was first reported in Japan over a century ago. A bacterial blight of rice was reported from Japan and from the Philippines almost sixty years ago. In Japan appeared as an endemic

26 Diseases of Field Crops and their Management

disease since 1884. In the 1960s, bacterial blight became prevalent in other ricegrowing regions of Asia with the introduction of high yielding varieties line TN 1 and IR 8, which were susceptible to the disease. The disease and causal organism were described in detail from Japan by Uyeda and Ishiyama in 1922, who named the organism *Pseudomonas oryzae.* Dowson (1949) named it *Xanthomonas oryzae.* Dye (1982) renamed it as *Xanthomonas campestris* pv. *oryvu.* Now it is called *X. oryzae* pv. *oryzae.* In India, it was reported by Srinivasan in Maharastra during 1959, but it receives more attention due to its epidemic outbroke in Bihar 1963. It epidemic was due to introduction of Taichung Native TN 1 variety.

Symptoms

1. **Kresek or wilt phase** (Reitsma and Schune, 1950):

» The disease is usually noticed at the time of heading but it can occur earlier also.

» Seedlings in the nursery show circular, yellow spots in the margin, which enlarge, coalesce leading to drying of foliage. "Kresek" symptom is seen in seedlings, 1-2 weeks after transplanting.

» The bacteria enter through the cut wounds in the leaf tips, become systemic and cause death of entire seedling.

2. **Leaf blight phase** (Gato, 1965):

» In grown up plants water soaked, translucent lesions appear near the leaf margin.

» The lesions enlarge both in length and width with a wavy margin and turn straw yellow within a few days, covering the entire leaf.

» As the disease advances, the lesions cover the entire lamina which turns white or straw coloured.

» Milky or opaque dew drops containing bacterial masses are formed on young lesions in the early morning.

» They dry up on the surface leaving a white encrustation.

» The affected grains have discoloured spots. If the cut end of leaf is dipped in water, it becomes turbid because of bacterial ooze.

3. **Yellow leaf or pale yellow phase**

» In the tropics, yellow leaf or pale yellow syndrome is associated with bacterial blight.

» The youngest leaf of the plant becomes uniformly pale yellow or has a broad yellow stripe.

Diseases of Rice

» With be yellow leaf, the bacteria are not present in the leaf itself but can be found in the internodes and crowns of affected stems.

Bacterium: *Xanthomonas oryzae* pv. *oryzae* (ex Ishiyama 1922, Swings *et al*, 1990)

The bacterium is aerobic, gram negative, non spore forming, rod with size ranging from 1-2 × 0.8-1.0 μm with monotrichous single polar flagellum.

Bacterial colonies are circular, convex with entire margins, whitish yellow to straw yellow colored (xanthomonadin) and opaque (extrapolysaccharide).

Disease cycle

Bacteria enter through hydathodes at the leaf tip or margin, then multiply in intercellular spaces and spread through the xylem. This is different to the invasion and development of *X. oryzae* pv. *oryzicola* which causes Bacterial Leaf Streak of rice. Access into the plant can also occur through wounds and other openings.

Bacteria move vertically and laterally along the veins and ooze out from the hydathodes, beading on the leaf surface. Wind and rain disseminate bacteria, monsoon season being the worst time for infection. Contaminated stubble, irrigation water, humans, insects and birds are also sources of infection. Bacteria can survive the winter even in temperate regions in weeds or in stubble. They can survive in soil for 1-3 months.

The infected seeds as a source of inoculum may not be important since the bacteria decrease rapidly and die in the course of seed soaking. The pathogen survives in soil and in the infected stubbles and on collateral hosts *Leersia* spp., *Plantago najor, Paspalum dictum*, and *Cyanodon dactylon.*

The pathogen spreads through irrigation water and also through rain storms. The mucous capsule is soluble in water, precipitated by acetone and is a polysaccharide, called Xanthomonadin.

Favorable conditions

» Clipping of tip of the seedling at the time of transplanting.

» Heavy rain, heavy dew, flooding, deep irrigation water.

» Severe wind and temperature of 25-30°C.

28   Diseases of Field Crops and their Management

» Application of excessive nitrogen, especially late top dressing.

Confirmation test for BB

» Cut the infected leaves and put them into test tube containing water. Yellowish bacterial ooze streams out from the cut ends into the water. After 1-2 hours, the water becomes turbid.

Economic Threshhold Level: Tillering to booting, flowering and after: 2-5% leaf damaged.

Management

Assay using ooze out test and nitrogen supplement seedling symptom test.

» Burn the stubbles.

» Use optimum dose of fertilizers.

» Split application of nitrogenous fertilizers.

» Avoid clipping of tip of seedling at the time of transplanting.

» Avoid flooded conditions.

» Remove weed hosts.

» Grow resistant varieties *O. longistaminata* and *O. rufipogan.*

» Grow resistant varieties ADT 36, ADT 39, IR 20, IR 22, IR 24, IR 28, IR 36, IR 54, IR 64, Ajaya, Asha, Biraj, Bharathidasan, Co 43, Co 45, CR 1094, MR 1550, MR 1523, Dharitri, Kunti, Lalat, Sarama, Suresh, Gobind, IR-64, Janaki, Karuna, Jawahar, Mandya Vijaya, Madhu, Indra, Mahsuri, Swarna, Saleem, Tholakari, Tikkara, Neerya, PR-4141, Pavizham, Radha, Ratna, Ruchi, Sona Mahsuri, Sujata, Suraj, Swarna, Sakti, Tetep, Udaya, Vaigai, and TKM 6.

» Seed and seedling bacterization with plant growth promoting rhizobacteria, *Azosprillum brasilense* and *Bacillus poymyxa* individually and in mixture.

» Seed treatment with *Bacillus subtilis, Pseudomonas fluorescens* and *Trichoderma harzianum* 2 g/ kg.

» Spray *Adhatoda vasica* was most effective followed by *Curcuma longa, Allium cepa, Prosophis juliflora* and *Azadirachta indica*

Diseases of Rice 29

» Seed treatment with bleaching powder (100g/ l) and zinc sulfate (2%) reduce bacterial blight.

» Seed treatment with copper hydroxide 0.1%, streptocycline 0.3 g and copper sulphate 1 g or streptocycline 0.3 g and copper oxychloride 1 g/ lit for 20 minutes.

» Seed soaking with Agrimycin-100 at 0.025% (15% streptomycin and 1.5% oxytetracyclin) followed by hot water dipping at 52°C - 54°C for 30 min.

» Steeping the seeds in mixed solution of tolyl mercury acetate (wettable ceresan at 500-1000 ppm) and agrimycin 100 (250 ppm) or streptocyclin (27 ppm) followed by hot water treatment at 52-55°C for 20 min.

» Spray Azadiractin (nemadol, neemzol, neemgold and tricure), neem oil 3% or NSKE 5% or neem soap oil.

» Spray fresh cowdung extract for the control of bacterial blight. Dissolve 20 g cowdung in one litre of water; allow settling and sieving.

» Spray streptomycin sulphate and tetracycline combination 300 g + copper oxychloride 1.25 kg/ ha.

» Application of bleaching powder at 5 kg/ ha in the irrigation water is recommended in the kresek stage.

» Foliar spray of streptocycline 0.05 g and copper sulphate 0.05 g.

» Two sprays of nanocopper at 0.2 ppm at fortnight interval from tillering stage.

» Spray synthetic bactericide such as nickel dimethyl dithiocarbamate, dithianone, phenazine and phenazine N oxide.

» Spray systemic resistance inducers *viz.*, benzothiadiazole, isonicotinic acid *etc.*,

16. Bacterial leaf streak

The disease was first described as stripe disease by Reinking 1918 from Phillipines. In India the disease was first reported in 1967 from Uttar Pradesh.

Symptoms

» Initially, small dark-green and water-soaked streaks on interveins from tillering to booting stage.

» The progress of the streaks is longitudinal, limited by the veins and soon turn yellow or orange brown.

» All along the streaks bacterial exudates could be observed as tiny yellow or amber colored droplets.

» These streaks may coalesce to form large patches and cover the entire leaf surface.

» Lesions turn brown to greyish white then dry when disease is severe Infection in the florets and seeds results in brown or black discoloration and death of ovary, stamens and endosperm and browning of glumes.

Bacterium: *Xanthomonas oryzae* pv. *oryzicola* (Fang *et al*. 1957) Swings *et al*. 1990.

The bacterium is a gramnegative, non-spore forming rod, $1.2 \times 0.3\text{-}0.5$ μm with a single polar flagellum.

Disease cycle

Bacteria survive from season to season in crop debris. Transmission occurs by seed in summer crops, and in irrigation water. Bacteria enter leaves through stomata and surface damage, often caused by insects. Masses of bacteria develop in the parenchyma.

Favorable conditions

» Presence of the bacteria on leaves and in the water or those surviving in the debris left after harvest.

» Rain and high humidity (83-93%) favors development of the disease.

» Warm temperature and high humidity

» Heavy wind

» Closer planting

» Frequent rainfall

» Early stage of planting from maximum tillering to panicle initiation

Management

» Burn the stubbles.

» Avoid clipping of tip of seedling at the time of transplanting.

Diseases of Rice

- » Avoid flooded conditions, remove weed hosts and use optimum dose of fertilizers.

- » Grow resistant varieties Bala, Padma, Sabarmathi, IR 20, Jamuna, Krishna, Jagannath and TKM 6.

- » Seed treatment i.e overnight soaking of seeds in 0.025% streptocycline solution and hot water treatment at 52°C for 30 minutes are effective in eradicating seed infection.

- » Spray neem oil 3% or NSKE 5% or neem soap oil.

- » Spray fresh cowdung water extract or lemon grass or mint extract at 20%.

- » Spray streptomycin sulphate and tetracycline combination 300g + copper oxychloride 1.25 kg/ha.

Differences between bacterial blight and bacterial streak

S. No	Bacterial blight	Bacterial streak
1.	The marginal lesion caused by *Xanthomonas oryzae* pv. *oryzae* are opaque against the light and lesions are formed at the leaf tip and margin.	The interveinal streaks caused by *Xanthomonas oryzae* pv. *oryzicola* are transparent.
2.	Lesion caused by *X. oryzae* pv. *oryzicola* pathogen remain wavy.	Streak/ Lesion margins caused by *X. oryzae* pv. *oryzicola* pathogen remain linear.
3.	*X. oryzae* pv. *oryzae* enters mainly through either wounds or hydathodes, multiplies in the parenchymatous cell and moves to the xylem vessels where active multiplication results in blight on the leaves.	*X. oryzae* pv. *oryzicola* enters mainly through the stomata and multiply in the parenchyma tissues of the leaves. It infects mainly the parenchymatous cells of the leaves, but is not systemic.

17. Rice tungro disease (RTD)

Rice tungro disease (RTD) is widely distributed in South and South-east Asian countries and has been recognised as a serious constraint to rice production. In the Filipino dialect, the word "Tungro" means degenerated growth and the disease.

Symptoms

- » Infection occurs both in the nursery and main field.

» Plants are markedly stunted.

» Leaves show yellow to orange discoloration and interveinal chlorosis.

» Young leaves are sometimes mottled while rusty spots appear on older leaves.

» Tillering is reduced with poor root system.

» Panicles not formed in very early infection, if formed, remain small with few, deformed and chaffy grains.

Virus

Two morphologically unrelated viruses present in phloem cells.

» *Rice tungro bacilliform virus* (RTBV) bacilliform capsid, circular ds DNA genome under *Tungrovirus* (*Caulimoviridae*) [disease causing virus] and

» *Rice tungro spherical virus* (RTSV) isometric capsid ss RNA genome under *Waikavirus* (*Sequiviridae*) [vector binding virus].

Vector: *Nephotettix virescens, Nephotettix nigripictus, Nephotettix apicalis and Racilia dorsalis.*

Iodine test

» Tungro infected plants can be identified by iodine test. Ten centimeter long leaf tip is cut in the early morning before 6 am and dipped in a solution containing 2 g iodine and 6 g potassium iodide in 100 ml of water for 30 minutes or dip in10 ml of tincture iodine + 140 ml of water for one hour. Tungro infected leaves show dark blue streaks.

Disease cycle

Transmission mainly by the leaf hopper vector *Nephotettix virescens* Males, females and nymphs of the insect can transmit the disease. Both the particles are transmitted semi-persistently; in the vector the particles are non-circulative and non-propagative. Plants infected with RTSV alone may be symptomless or exhibit only mild stunting. RTBV enhances the symptoms caused by RTSV. RTSV can be acquired from the infected plant independently of RTBV, but acquisition of RTBV is dependent on RTSV which acts as a helper virus. Both the viruses thrive in rice and several weed hosts which serve as source of inoculum for the next. Ratoon from infected rice stubble serve as reservoirs of the virus. Disease incidence depends on rice varieties, time of planting, time of infection and presence of vectors and favorable weather conditions,

Diseases of Rice 33

Management

» Field sanitation, removal of weed hosts of the virus and vectors.

» Grow disease tolerant varieties like Pusa 2-21, IR 20, IR 26, IR 36, IR 50, ADT 37, ASD 17, ASD 18, Saket-4, PTB 18, Ponmani, Co 45, Co 48, CoRH 1, CoRH 2, MTU 9992, MTU 1002, MTU 1003, MTU 1005, IET 2815, Rashi, IET 2395, CWM 31, Aravinder, Kunti, Puduvaiponni, Punithavathy, Bharathidasan, Bharani, Hira, Vasundara, Srinivas, Surekha, Suraksha, Vikramarya, Bharani and White ponni.

» Apply neem cake at 12.5 kg/ 20 cent or carbofuran 3 g/ sq.m nursery as basal dose.

» Leaf yellowing can be minimized by spraying urea 2% + mancozeb at 2.5 gm/ lit.

» Instead of urea, foliar fertilizer like multi-K (potassium nitrate) can be sprayed at 1%.

» Set up light traps to monitor the vector population.

» Spray *Paecilomyces* sp. 10g/ lit. of water

» The vegetation on the bunds should also be sprayed with the insecticides. Maintain 2.5 cm of water in the nursery and broadcast anyone of the following in 20 cents carbofuran 3 G 3.5 kg or phorate 10 G 1.0 kg or quinalphos 5 G 2.0 kg.

» During pre-tillering to mid-tillering when one affected hill/ m is observed apply phorate 10 G at 12.5 kg/ ha or carbofuran 3 G at 17.5 kg/ ha or Quinalphos 5 G at 20 kg/ ha control the vector in the main field 15 and 30 days after transplanting.

» Spaying of acephate 75% SP or chlorantraniliprole 0.5% + thiamethoxam 1.0% GR or imidacloprid 70% WG or thiamethoxam 25% WG or ethiprole 40 + imidacloprid 40% or acephate 75% WP 0.75 g/ lit or quinalphos 25% EC 1 ml/ lit phosphomidon 40% SL at 1.5 ml/ lit.

18. Rice grassy stunt disease

Rice grassy stunt disease described by Rivera *et al*. (1966) and Bergonia *et al*. (1966) as Rice rosette virus. Virus characterized by Hibino *et al*. (1985).

Symptoms

» Plants are markedly stunted with excessive tillering and an erect growth habit.

» Leaves become narrow, pale green with small rusty spots and plant becomes grassy or rosette appearance.

34 Diseases of Field Crops and their Management

> May produce a few small panicles which bear dark brown unfilled grains.

Virus: *Rice grassy stunt virus*, (*Tenuivirus*) flexuous, filamentous 950-1350 nm long × 6 nm wide, ssRNA genome.

Disease cycle

Disease spreads by the brown plant hopper, *Nilaparvata lugens,* in a persistent manner having a latent period of 5 to 28 days in the vector. Ratoon crop and presence of vector perpetuate the disease from one crop to other.

Management

> Avoid close planting and provide 30 cm rogue spacing at every 2.5-3.0 m to reduce the pest incidence.

> There are varieties released by IRRI, which contain genes for BPH resistance, like IR 26, IR 28, IR 29, IR 30, IR 32, IR 34, IR 36, IR 56, IR 64 and IR 72.

> Plouging and fallowing the field to destroy stubbles right after harvest in order to eradicate other hosts.

> Spray phosphamidon 40 SL 1000 ml/ ha or phosalone 35 EC 1500 ml/ ha or carbaryl 10 D 25 kg/ha or acephate 75 SP 625 gm/ ha or chlorpyriphos 20 EC 1250 ml/ ha or Acetamiprid 20% SP or Benfuracarb 3% GR or Bifenthrin 10% EC or Clothianidin 50% WDG or Dinotefuran 20% SG or Fipronil 5% SC or Flonicamid 50% WG or Imidaclopride 70% WG or Thiamethoxam 25% WG or Ethiprole 40 + Imidacloprid 40%WG.

19.   Rice ragged stunt disease

Rice ragged stunt virus (RRSV) disease was first observed in the Philippines and identified by Ling *et al*. Before this, de la Curz and Hibino *et al* described identical symptomatology and termed the diseases "infectious galls" and "Kerdil hampa," respectively. The disease was reported in India by Ghosh *et al.*

Symptoms

> Formation of ragged leaves with irregular margins, vein swelling, and enations on leaf veins may be formed.

Diseases of Rice

- » Leaves short and dark green with serrated edges, ragged portions of the leaves are yellow to yellow-brown and twisted into spiral shapes at the base of leaf blades swollen, pale yellow or white to dark brown veins developing on leaf blades and sheaths galls on the underside of leaf blades and outer surface of leaf sheaths twisted, malformed flag leaves that are shortened at booting stage.

- » Stunting of plants at early growth stage, delayed flowering, production of nodal branches and incomplete emergence of panicles.

- » The ragged appearance and twisted leaf symptoms can be confused with the damage caused by rice whorl maggot and nematodes.

- » To confirm rice grassy stunt check for the presence of the brown plant hopper vector, vein swelling and dark green color of leaves as well as severe stunting.

Virus: *Rice ragged stunt virus* (*Figivirus*), Spherical virus, 65 nm diameter, dsRNA genome.

Disease cycle

Spreads through brown planthopper, *Nilaparvata lugens* transmitted in a persistent manner. Multiplies in the vector, latent period of 3-35 days, but not transmitted congenitally

Management: Same as that of Rice grassy stunt disease

20. Rice dwarf disease

Symptoms

- » Infected plants show stunted growth, reduced tillering and root system.

- » Leaves show chlorotic specks turning to streaks along the veins.

- » In early stage of infection no ear heads formed.

Virus: *Rice dwarf virus*

- » The virus is spherical, 70nm diameter with an envelope, dsRNA genome.

Disease cycle

Leafhoppers (*Nephotettix cincticeps, Recillia dorsalis* and *N. nigropictus*) transmit in a persistent manner. The transmission is transovarial through eggs. Gramineous weeds

Echinochloa crusgalli and *Panicum miliaceaaum* serve as source of inoculum.

Management

- » Destory weed host that serve as source of inoculum

- » Spray phosphamidon or fenthinon 500 ml or monocrotophos 1 lit/ ha.

21. White leaf disease (*Rice hoja blanca virus*) - transmitted by *Sogatella cubana*

Infected plants show leaf mottling and complete chlorosis. Severely infected plants dry prematurely without forming grains. The spikes become white and are distributed in the field. Similar white spikes may also be due to the attack of the stem borer *Diatrea saccharalis* Fabr., or because of frost. The Rice hoja blanca can be distinguished from the other causes, since *D. saccharalis* creates a small hole at the base of the stem which can be easily observed and in the case of frost damage, the spikes can be easily removed by fingers from the stem. This disease also reported in Wheat.

Other virus diseases

Rice transistary yellow virus (Suffocation disease) - by *Nephotettix nigropictus*
Rice stripe tenuivirus

22. Rice yellow dwarf disease

First documented in Kochi Agricultural Experiment Station by (Muniyappa and Rayachaudhuri, 1988). Yellow dwarf disease of rice has been reported in most of the rice growing countries of the East and South East Asia. In some areas of Japan and India, the disease has been observed to cause serious damage to the crop (Hayashi, 1961; Raychaudhuri and Nariani, 1977).

Symptoms

- » Prominent stunting of plants and excessive tillering are the characteristic symptoms of the disease.

- » Leaves show yellowish green to whitish green, become soft and droop.

- » Plants usually remain sterile but sometimes may produce small panicles with unfilled grains.

Bacterium: *Candidatus* Phytoplasma oryzae

Diseases of Rice 37

Disease cycle

The disease is transmitted by leafhopper vectors *Nephotettix virescens* and *N. nigropictus* with a latent period of 25-30 days in the vector. The pathogen survives on several grass weeds.

Management

» Deep ploughing during summer months and burning of stubbles.

» Rice varieties IR 62 and IR 64 are moderately resistant to the disease.

» The management practices followed for Rice Tungro disease holds good for this disease also.

» Spraying of streptomycin sulphate WP 200 ppm minimize the disease followed spraying of magnesium silicate impart resistant against disease.

Difference between Tungro and Yellow dwarf disease

Tungro	Yellow dwarf
It was caused by Virus	It was caused by Phytoplasma
Symtoms observed during nursery and mainfield	Symtoms observed only in transplanted mainfield
Yellow or orange yellow lines alternative with dark green lines appear on leaves	Entire leaf become yellow
Reduced numbers of tillers	Increased numbers of tillers
Leaves stand erect	Leaves become soft and droop
Moderate stunting	Severe stunting
Virus lost their virulence in vector within 2 days	Phytoplasma retained in the leaf hopper in persistant manner

23. Ufra disease: *Ditylenchus angustus* (Butler) Filipjev.

» During vegetative growth from seedling to flag leaf, the principal symptom of infection is leaf chlorosis.

» In light infections, the chlorosis will be discrete white spots, less than 1 mm in diameter.

» *D. angustus* does feed on the inner surface of the leaf sheaths, but these rarely show obvious symptoms.

> » In time the chlorotic areas will show some localized browning.

> » Depending on the severity of infection, chlorotic leaf areas, tillers or whole plants will wither and die, attaining a light-brown appearance.

Other nematode: Rice white tip nematode – *Aphelenchoides besseyi*

24. Khaira disease: Zinc deficiency in calcaroius soil

Dr. Yashwant Laxman Nene and his associates first discovered Khiara disease of rice in Tarai region of the then Uttar Pradesh (Present Uttranchal).

> » Appears in small patches in nursery or in the main field.

> » In main field appearance of chlorotic bases of leaves at 10-15 days after transplantation and small bronze colored spots on leaves which later spread on entire leaf blade which becomes bronze colored and ultimately dried.

> » The infected plants show stunted growth and no ear formation on diseased plants or if formed no grains.

Management

> » Foliar spray with zinc sulphate 5 kg + slaked lime 2.5 kg in 1000 litres/ ha, first spray 10 days after transplant (DAS) in nursery; second spray at 20 DAS and third-within 15-30 DAS.

> » Apply zinc sulphate 25 kg/ ha along with phosphatic fertilizers

Chapter - 2

Diseases of Wheat - *Triticum aestivum* L.

S. No.	Disease	Pathogen
1.	Black (stem) rust	*Puccinia graminis* var. *tritici*
2.	Orange brown (leaf) rust	*Puccinia triticina*
3.	Yellow (stripe) rust	*Puccinia striiformis*
4.	Loose smut	*Ustilago tritici*
5.	Flag smut	*Urocystis tritici*
6.	Rough spored bunt	*Tilletia tritici*
7.	Smooth spored bunt	*Tilletia laevis*
8.	Dwarf bunt	*Tilletia controversa*
9.	Karnal bunt	*Tilletia indica*
10.	Powdery mildew	*Blumeria graminis* f.sp. *tritici*
11.	Leaf blight	*Alternaria triticina*
12.	*Pythium* foot rot	*Globisporangium abappressorium*
13.	Pink snow mold	*Microdochium nivale*
14.	*Fusarium* foot rot	*Fusarium culmorum*
15.	Tan spot	*Pyrenophora tritici-repentis*
16.	Tundu	*Clavibacter tritici* + *Anguina tritici*

17.	Bacterial leaf streak and black chaff	*Xanthomonas translucens*
18.	Soil borne mosaic disease	*Soil-borne mosaic virus*
19.	Take all disease	*Gaeumannomyces graminis*
20.	Scab (The Fusarium Head Blight)	*Fusarium graminearum*
21.	Leaf blotch	*Mycosphaerella graminicola; Leptosphaeria nodorum; Leptosphaeria avenaria*
22.	Molya disease	*Heterodera avenae*

Wheat Rust

Aristotle (384-322 B.C.) writes of rust being produced by the "warm vapors" and mentions the devastation of rust and years when rust epidemics took place. Theophrastus reported that rust was more severe on cereals than legumes.

Wheat rust pathogens belong to genus *Puccinia*, family Pucciniaceae, order Uredinales and class Basidiomycetes. These rust fungi are highly specialized plant pathogens with narrow host ranges. The Italians Fontana and Tozzetti independently provided the first unequivocal and detailed reports of wheat stem rust in 1767 (Fontana, 1932; Tozzetti, 1952). Chester (1946) provided one of the first detailed histories of the literature on the rust of wheat. In the early records, wheat leaf rust is not distinguished from stem rust (Chester, 1946).

The first stem rust epidemic record goes back to 1786 A.D. in central India. Widespread occurrence of leaf rust was observed during 1971-73 in popular cultivar Kalyansona in northern plains. Both leaf rust and stripe rust occurred each year from 1967 to 1974 but the losses were estimated only twice.

1. Black (stem) rust

Stem rust, also known as black rust, is one of the most studied rusts of wheat in the world and occurs wherever the crop is grown. In this rust, stems are more severely attacked than the other parts of the plant like leaves, sheaths and awns and hence the names "stem rust". The causal organism of wheat stem rust was named *P. graminis* by Persoon in 1797.

In 1999, a new virulent race of stem rust was identified from wheat fields in Uganda – popularly known as Ug99 after the year and country of discovery. Using North American

Diseases of Wheat 41

scientific nomenclature, Ug99 is known as race TTKSK. Ug99 (Race TTKSK) is a cause for concern as it exhibits unique virulence patterns. No other race of stem rust has been observed to overcome so many wheat resistance genes, including the very important gene *Sr31*. During 2007, Ug99 (Race TTKSK) had spread via wind movements out of East Africa, into Yemen and as far as Iran. There is now 11 known races of Ug99, all are closely related or evolved from common ancestor.

Symptoms

» Symptoms are produced on almost all aerial parts of the wheat plant but are most common on stem, leaf sheaths and upper and lower leaf surfaces.

» Uredial pustules (or sori) are oval to spindle shaped and dark reddish brown (rust) in color.

» They erupt through the epidermis of the host and are surrounded by tattered host tissue. The pustules are dusty in appearance due to the vast number of spores produced. Spores are readily released when touched.

» As the infection advances teliospores are produced in the same pustule. The color of the pustule changes from rust color to black as teliospore production progresses.

» If a large number of pustules are produced, stems become weakened and lodge. The pathogen attacks other host (barberry) to complete its life cycle.

» Symptoms are very different on this woody host. Other spores are Pycnia (spermagonia) produced on the upper leaf surface of barberry which appears as raised orange spots.

» Small amounts of honeydew that attracts insects are produced in this structure. Aecia, produced on the lower leaf surface, are yellow.

» They are bell-shaped and extend as far as 5 mm from the leaf surface.

Fungus: *Puccinia graminis* Pers. var. *tritici* Erikss. & E. Henning

The fungus is an obligate biotrophic parasite. It possesses polymorphic, heterocious nature. Fontana (1932) made the first known detailed study, including precise drawings, of *Puccinia graminis* in 1767. Persoon named the fungus on barberry *Aecidium berberidis* in 1791 and the form on wheat *Puccinia graminis* in 1794. De Bary (1866) showed that the two fungi were different stages of a single species. Craigie (1927) made the first controlled crosses between strains of *Puccinia graminis*.

Erikson (1894) reported physiological specialization of *Puccinia graminis*

»   *Puccinia graminis* var. *tritici* infects wheat, barley and some grasses

»   *Puccinia graminis* var. *secalis* infects rye, barley and some grasses

»   *Puccinia graminis* var. *avenae* infects oat and some wild grasses

»   *Puccinia graminis* var. *phleipratensis* infects timothy grass and some wild grasses

»   *Puccinia graminis* var. *agrostidis* infects *Agrostis* spp.

»   *Puccinia graminis* var. *poae* infects *Poa pratensis* and related species and kentucky blue grass.

Over 250 physiologic races of *P. graminis tritici* have been differentiated so far. While many of them are extinct or present only in insignificant proportions, a few are widespread in highly virulent forms. In India, 26 races (biotypes) have been reported.

Stages

Stage 0	Pycnium bearing spermatia (Pycniospore) or receptive hyphae (n) or (n')	Plasmogamy (Monokaryotic mycelium)	Survives in Barberry - Dicot (alternate host)
Stage I	Aecium bearing aeciospores $(n+n')$ (haploid)	Dikaryophase starts	
Stage II	Uredinium bearing Urediniospores $(n+n')$	Dikaryotic mycelium	Survives in Wheat - Monocot (main host)
Stage III	Telium bearing Teliospores $(n+n'-\!\!\gg 2n)$	Karyogamy (Diplophase)	
Stage IV	Basidium produces promycelium, which bearing Basidiospores $(n)+(n')$	Germination meiosis (Diplophase)	

Disease cycle

In India, all these rusts appear in wheat growing belt during Rabi crop season. Uredosori turn into teliosori as summer approaches. The inoculum survives in the form of uredospores / teliospores in the hills during off season on self sown crop or volunteer hosts, which provide an excellent source of inoculum. Alternate hosts are *Barberis vulgaris* L., *B. canadensis* Mill., *B. fendleri* and *Mahonia* sp in the world. In India, role of alternate host (Barberis) is not there in completing the life cycle.

Diseases of Wheat 43

The fungus is inhibited by temperatures over 20°C although strains tolerant of high temperatures do exist. The complete cycle from infection to the production of new spores can take as little as 7 days during ideal conditions. The disease cycle may therefore be repeated many times in one season. During late summer, the dark teliospores may be produced. These can germinate to produce yet another spore type, the basidiospore, but no alternate host has been found. Although the teliospores seem to have no function in the disease cycle they may contribute to the development of new races through sexual recombination.

Uredospores and dormant mycelium survive on stubbles and straws and also on weed hosts and self sown wheat crops. Wind borne uredospores from hills are lifted due to cyclonic winds and infect the crop in the plains during crop season.

Favourable conditions

» Low temperature (15-20°C) and high humidity during March-April in Northern India and November –December in Southern India.

» Temperature less < 10°C favours rust disease.

Management

» Grow resistant varieties like Lerma roja, Girija, Sonalika, Chotilerma, Safed lerma, Sonara 64, S-227, S-307, HD-2278, WL-359, C-306, HW-741, NP 700, NP 800, PBW 343, PBW 550, and PBW 17.

» Mixed cropping with suitable crops.

» Avoid excess dose of nitrogenous fertilizers.

» Spray lime-sulphur, hydrophobic colloidal sulphur, nickel chloride, nickel sulphate and dithiocarbamates control this and other rust diseases.

» Spray propiconazole 25% EC at 1 ml/ lit of water or tebuconazole 25% EC at 1 ml/ lit or zineb 75% WP at 2 g/ lit or mancozeb 75% WP at 1.5 g/ lit.

» Combined spraying of prothioconazole 2 g/ lit + tebuconazole 2 g/ lit or propiconazole 2.5 g/ lit + cyproconazole 0.8 g/ lit.

» Spray parzate liquid and zinc sulphate at three to four times.

» Spray with cyclohexamide (actidione) at 50 ppm thrice.

2. Orange brown (leaf) rust

In 1815 de Candolle had shown that wheat leaf rust was caused by a distinct fungus *Uredo rubigovera*. The pathogen underwent a number of name changes until 1956 when Cummins and Cald-well (1956) suggested *P. recondita*, which has been the generally used nomenclature.

Symptoms

> » The most common site for symptoms is on leaf blades, however, sheaths, glumes and awns may occasionally become infected and exhibit symptoms.

> » Uredia are seen as small, circular orange blisters or pustules on the upper surface of leaves.

> » Orange spores are easily dislodged and may cover clothing, hands or implements. When the infection is severe leaves dry out and die. Since inoculum is blown into a given area, symptoms are often seen on upper leaves first.

> » As plants mature, the orange urediospores are replaced by black teliospores.

> » Pustules containing these spores are black and shiny since the epidermis does not rupture.

> » Yield loss often occurs as a result of infection by *Puccinia recondita* f. sp. *tritici.*

> » Heavy infection which extends to the flag leaf results in a shorter period of grain fill and small kernels.

Fungus: *Puccinia triticina* Erikss.
Syn: *Puccinia recondita* Rob. ex Desm. f. sp. *tritici* (Erikss.),
P. rubigo-vera tritici (Ericks.). Carleton, *P. rubigo-vera* (DC) Wint.

The uredinospores are brown, round and measure 16-20 µm in diameter. On germination they form an appressorium and infect the plant through stomata. The teliutospores are rarely produced, dark brown with 2-3 septa and are similar to those of yellow rust fungus.

Disease cycle

The alternate hosts are: *Thalictrum flavum* L., in India; *Isopyron fumarioides* W., in Russia; *Thalictrum foetidum* L., *Thalictrum japonicum* Thumb. *Anchusa italica* Retz., *Clematis mandscurica* Rupr., and *Thalictrum speciosissimum* Loeft., in Asia, America and Mediterrenia.

Diseases of Wheat 45

The collateral hosts are: *Briza minor, Bromus pamlus. Brachypodium sylvaticum* and *Avena fatua* are the grasses that become naturally infected.

» Brown rust overwinters in crops and on volunteers.

» It spreads by airborne spores. Cold winters may reduce its survival.

Favourable conditions

» Optimum conditions are days with high temperatures (15-22°C) followed by overnight dews.

» Surface moisture on leaves is essential for spore germination.

» Symptoms can appear in 5–6 days at optimum temperatures.

» The disease is active over a wider range of temperature (7-25°C) than yellow rust.

Management

» Genetic resistance is the best way to control the rust, while specific resistance genes like *Lr 19, Lr22, Lr29, Lr32* and *Lr 33* were effective for several years.

» Grow resistant varieties like Lerma roja, Sonalika, Shailya, Janak, Arjun, Malabika, UP 262 and HB 208.

» Mixed cropping with suitable crops.

» Avoid excess dose of nitrogenous fertilizers.

» The parasite *Darluca filum (Sphaerellopsis filum)* Cast., imperfect stage of *Eudarlucca cacicis* (Fr.) Ericks., attacks the rust under natural conditions.

» Spraying of mycoparasite *Sphaerellopsis filum* and *Beauveria bassiana* immediately after development of symptom.

» Propalytic spraying of *Trichoderma harzianum, Bacillus subtilis* and yeast *Saccharomyces cerevisiae* are effective against wheat rust.

» Spray with fluxapyroxad + pyraclostrobin 300 ml/ ha.

» Spraying of flutriafol 0.5 ml/ lit or propiconazole 1 ml / lit or tebuconazole 2 g/ lit or bitertanol 2 g/ lit.

» Combined spraying of prothioconazole 1 g/ lit + tebuconazole 2 g/ lit or azoxystrobin

46   Diseases of Field Crops and their Management

1 g/ lit + cyproconazole 0.4 g/ lit or propiconazole 1 g/ lit + cyproconazole 0.4 g/ lit.

3. Yellow (stripe) rust

Recent morphological studies by Savile (1984) and morphological and pathogen genetic studies by Anikster *et al*. (1997) show that *P. recondita* is not the incitant of wheat leaf rust. Currently *P. triticina* should be the preferred name as shown in by Savile (1984) and Anikster *et al*. (1997). This name was used by Mains and Jackson (1926) and has been used in parts of Asia and Eastern Europe for many years. Although Gadd first described stripe rust of wheat in 1777, it was not until 1896 that Eriksson and Henning (1896) showed that stripe rust resulted from a separate pathogen, which they named *P. glumarum*. In 1953, Hylander *et al*. (1953) revived the name *P. striiformis*.

Symptoms

    »   Mainly occur on leaves than the leaf sheaths and stem.

    »   Bright yellow pustules (Uredia) appear on leaves at early stage of crop and pustules are arranged in linear rows as stripes.

    »   The stripes are yellow to orange yellow.

    »   The teliospores are also arranged in long stripes and are dull black in colour.

Fungus: *Puccinia striiformis* Westend.
         Syn: *Puccinia glumarum* Erikss. & E. Henn., *Puccinia striiformis* f. sp. *tritici* West.

The uredospores of rust pathogen are almost round or oval in shape and bright orange in colour. The teliospores are bright organge to dark brown, two celled and flattened at the top. Sterile paraphyses are also present at the end of sorus. The alternate hosts are suspected as *Agropyron* sp. and *Bromes* sp in foreign.

Some weeds like *Agropyron spicatum, Bromus catharticus, B. japonicas* and *Hordeum murinum* and over 90 other grasses also are susceptible to stripe rust.

Management

Stripe rust is controlled by a combination of genetic resistance and fungicide applications. The slow rusting varieties developed over the past 50 years which include *Lr34/Yr18* and *Lr46/Yr29* and Pavon 76.

Diseases of Wheat

» Mixed cropping with suitable crops.

» Avoid excess dose of nitrogenous fertilizers.

» Grow resistant varieties like DWL 5023, HD 2281, UP 262, WH 47, Lerma roja and Sonalika.

» Spray mycoparasite *Sphaerellopsis filum* and *Beauveria bassiana* immediately after development of initial symptom.

» Spraying of flutriafol 0.5 ml/ lit or propiconazole 1 ml/ lit or Triadimefon 0.5 g/ lit or tebuconazole 1 ml/ lit.

» Combined spraying of azoxystrobin 1 g/ lit + cyproconazole 0.4 g/ lit or propiconazole 1 g/ lit + cyproconazole 0.4 g/ lit or prothioconazole 1 g/ lit + tebuconazole 2 g/ lit.

Differences within the wheat rust diseases

Disease	Stem rust	Leaf rust	Stripe rust
Symptoms	Random, oblong pustules with torn margins.	Random, circular to oval pustules.	Small closely packed circular pustules during the vegetative stage, becoming stripes along leaves of older plants.
Plant part affected	Both sides of leaf, leaf sheaths, stems and outside of head	Upper surface of leaf and leaf sheaths	Upper surface of leaf, leaf sheaths, awns and inside glumes
Spore colour	reddish/ black	orange/ brown	yellow/ orange

Smut

The first written account of the cereal smuts comes from Theophrastus (384-332 BC). Smut was known to the Romans, who named it Ustilago, which comes from the Latin word for burn. This term was later used in many languages as the common name for smut (e.g., carbon, carbone, sot, and brand). The rust fungi produce fi ve types of spores whereas the smut fungi produce only two types of spores: the clamydospores (teliospores) and the basidiospores (sporidia).

4. Loose smut

Loose smut (LS) of wheat was illustrated in 1556 in Hieronymus Bock's Herbal, and an accurate symptomology is given in Fabricius' text of 1774.

Symptoms

» It is very difficult to detect infected plants in the field until heading, at this time; infected heads emerge earlier than normal heads.

» The entire inflorescence is commonly affected and appears as a mass of olive-black spores, initially covered by a thin gray membrane.

» Once the membrane ruptures, the head appears powdery.

» Spores are dislodged, leaving only the rachis intact.

» In some cases remnants of glumes and awns may be present on the exposed rachis.

» Smutted heads are shorter than healthy heads due to a reduction in the length of the rachis and peduncle.

» Infected heads are shorter; the rest of the plant is slightly taller than healthy plants. Prior to heading affected plants have dark green erect leaves.

» Chlorotic streaks may also be visible on the leaves. All the tillers of infected plant show infected spikes in which no grains are formed.

Fungus: *Ustilago tritici* (Pers.) Rostr.
      Syn: *Ustilago nuda* var. *tritici* G.W. Fisch. & C.G. Shaw and *U. nuda* (Jens.) Rostr.

The teliospores are black, finely raticulated and measure 5–8 μm in diameter. Loose smut is a systemic disease and is **internally seed-borne**. The infection occurs at the time of flowering. The teliospores are blown by the wind, get deposited on the flowers of wheat spike where they germinate and form promycelium. The promycelia fuse and produce infective hyphae which penetrate through the ovary wall and thus causes seed infection. After infection the mycelium stays dormant in the embryo until the germination of the seed.

Disease cycle

Ears of infected plants emerge early. The spores released from the infected heads land on the later emerging florets and infect the developing seed. Infection during flowering is favored

Diseases of Wheat 49

by frequent rain showers, high humidity and temperature. The disease is internally seed borne, where pathogen infects the embryo in the seed.

Management

» Grow resistant varieties like Kalyan Sona, Sonalika, Chotilerma, Kalyan 227, PV 18, WG 307, NP 13, UP 262, UP 718, UP 768, IP 780, HD 1950 and C 302.

» Hot water treatment (Jenson, 1886): Soak the seeds in water at 26-30°C for 5 hrs to induce dormant mycelium to grow. Then immerse the seeds in hot water at 54°C (129°F) for 10 min to kill the mycelium.

» Solar treatment (Luthra and Sattar, 1954 in Punjab): Soak the seeds in cold water for 4 hours from 6 AM to 10 AM in the forenoon on a bright sunny day followed by spreading and drying of seeds on brick floor in bright sun (44°C) for 4 hours from 10 AM to 5 PM in the afternoon.

» Soak the seeds in 20°C for 5 hours, draining for 1 minute, dipping in 49°C for 1 min and then at 52°C for 11 min and immediately put in cool water, then dry shade before sowing.

» Treat the seed with carboxin 2.5 g/ lit, chloronil 2 g/ lit or triadimenol at 2 g/ kg or mycobutanil 0.5 g/ lit seed before sowing.

» Burry the infected ear heads in the soil, so that secondary spread is avoided.

» Spray fungicides like carboxin 0.125% or tebuconazole 0.2%.

» Managing smut and bunt, general recommented fungicides: Metalaxyl, mefenoxam, triticonozole, difenoconozole, thiobendazole, captan, azoxystrobin, fludioxonil and mancozeb.

5.   Flag smut

Historically flag smut, caused by the fungus *Usrocystis agropyri*, was an important disease of wheat in Victoria. Flag smut of wheat was first reported in South Australia in 1968. The *U. agropyri*, is a basidiomycete. It produces basidiospores and teliospores. This pathogen is found globally, but is most problematic in Australia and India. *Urocystis agropyri* was first described from *Elymus repens* in Germany and *U. tritici* was first described from *Triticum vulgare* (=*T. aestivum*). In 1953, G. W. Fischer placed *U. tritici* and a large number of other *Urocystis* species in synonymy with *U. agropyri*.

Symptoms

» Flag smut is a systemic disease that starts in young tissues.

» Early symptoms include "leprous" spots and bending or twisting of coleoptiles.

» Flag smut is typically leaf and vein disease. Infected leaves and seedlings are twisted and become most conspicuous during stem elongation and early boot stage.

» The symptoms can be seen on stem, culm and leaves from late seedling stage to maturity.

» The seedling infection leads to twisting and drooping of leaves followed by withering.

» Grey to grayish black sori occurs on leaf blade and sheath.

» The sorus contains black powdery mass of spores.

» In the field infection can be distinguished by plants showing stunted, wilted and yellowish-green leaves.

Fungus: *Urocystis tritici* Korn 1877 or
Urocystis agropyri (Preuss) A.A. Fisch. Waldh., (1867)

Aggregated spore balls, consisting 1-6 bright globose, brown smoth walled spores surrounded by a layer of flat sterile cells. Teliospores are globose to subglobose.

Disease cycle

During harvest the black spores are released from the plant contaminating seed and soil. Typically spores survive in soil for 3 years, but can survive for up to 7 years.

Soil or seed borne spores infect the new wheat plant before emergence. Infection is favoured by early sowing into relatively dry and warm soils. Optimal temperature for infection is 20°C, but infection may occur at as low as 5°C and as high as 28°C. The fungus grows inter and intra cellularly between vascular bundles of the leaf tissue and other effected plant parts. Smut spores are viable for more than 10 years.

Favourable conditions

» Temperature of 18-24°C.

» Relative humidity 65% and above.

Diseases of Wheat

Management

» Grow resistant varieties like Sonalika, Sarbati Sonara, Pusa 44 and WG 377.

» Seed treatment with brine solution.

» Seed treatment with carboxin or tebuconazole or triademefon at 2g /kg.

» Dry seed-dressing with non-systemics such as copper carbonate, and systemics such as carboxin, oxycarboxin, pyracarbolid, fenfuram, triadimefon, triadimenol and tebuconazole.

Difference between Rust and Smut

Characters	Rust	Smut
Name and General chracters	Common name given because of the "rusty" appearance caused by disease, in urediospore stage. Many species parasic on grain crops. Characterized by lack of fruing bodies, formation of basidia and basidiospores from germinaon of teliospores and having as many as five spore stages (Polymorphic) and two hosts, e.g., *Puccinia graminis* (Wheat Rust)	Common name given because of the black, powdery appearance of infected host plants, in teliospore stage. Many species parasic on grain crops.
Parasitism	Wheat rusts are heteroecious and others are autoecious parasite	All smuts are autoecious saprophyte
Host range	Mostly angiosperms	Angiosperms, gymnosperms and ferns
Mycelium	The rusts are intecellular and obtain their nutrion by means of haustoria.	The smuts may be intercellular or intracellular (*U. maydis*). Haustoria are present.
Clamp connections	Clamp on the secondary mycelium are rare.	Clamp connections are common.

Spores	The dikaryotic mycelium produces three kinds of binucleated spores; uredospores and teleutospores on the primary host and aeciospores on the alternate host	It produces only one kind of binucleate spores called the smut pores which are comparable to the teleutospores of rusts.
Spore development	The teleutospores are developed from the terminal cells of the mycelium.	Smut spores are formed from the intercalary cells.
teleutospores	The teleutospores are stalked, two celled and each cell is binuclate.	The brand spores (teleutospores) are uninuclear and binucleate.
Germination of teleutospores	Each cell of the two celled teleutospores produces an epibasidium which bears four basidiospores. They are borne on sterigmata and are discharged violently by the water drop method.	The single celled, brand spore which is equivalent to teleutospores, germinates to produce a single epibasidium which bears a varible number of basidiospores. They are not borne on the sterigmata nor are they discharge violently.
Infection	Mostly systemic; Few localised	Localized; Never systemic

Common bunts

The main symptoms caused by these three species are fungal structures called "bunt balls," which resemble kernels but are completely filled with black teliospores. The fungus attacks seedling of 8-10 days old and become systemic and grows along the tip of shoot. At the time of flowering hyphae concentrate in the inflorescence and spikelets and transforming the ovary into smut sorus of dark green color with masses of chlamydospores. The diseased plants mature earlier and all the spikelets are affected.

Stinking smut, with its obvious masses of black, smelly teliospores, was noted long before the germ theory of disease was accepted.

The name "smut," which is derived from the Germanic word for "dirty," comes from this black spore stage. The name "bunt" comes from a dialectic contraction of the term "burnt ears" to "bunt ear" and finally to just "bunt."

Diseases of Wheat

Types

6. Rough spored bunt (European bunt/ hill bunt/ stinking smut)

- *Tilletia tritici* (Bjerk.) Syn: *Tilletia caries* (DC.) Tul. & C. Tul.

7. Smooth spored bunt

- *Tilletia laevis* J.G. Kühn Syn: *Tilletia foetida* (Wallr.)

8. Dwarf bunt (short smut / stunt smut / stubble smut / TCK smut)

- *Tilletia controversa* J.G. Kühn

The bunt balls of common bunt, caused by *T. caries* and *T. foetida*, are about the same size and shape as the kernels they replace; those of dwarf bunt, caused by *T. controversa*, are more nearly spherical.

Symptoms

» Infected wheat plants often, tend to be slightly shorter than healthy plants.

» After heading, the spikelets of infected plants tend to "flare-out" and take on a greasy, off-green color.

» This "flaring out" of the spikelet is due to the expansion in size of the bunt infected seed that has become filled with teliospores.

» In varieties that normally produce long awns (bristle-like structures), infected heads may have shorter awns, or even no awns.

» In place of normal seeds, infected kernels develop into "bunt balls".

» These are the remnants of what would normally be a seed, but in its place, the seed coat remains intact with the inside converted into a black mass of spores.

» Infected ovaries appear greasy with a dark green cast. When squeezed, such ovaries reveal a mass of black spores that smell like rotting fish. This odor is actually that of trimethylamine, which is produced by the smut fungus.

» As the heads and kernels mature, the bunt balls develop into a hardened mass that looks like miniature footballs.

»   The spores inside the mature bunt balls are released when the heads go through the combine harvester to produce the cloud of dust.

»   This dust also smells of rotting fish. Occasionally, both healthy seeds and bunt balls are found in the same head.

Fungal characters

The genus *Tilletia* was named by the Tulasne brothers in 1847 to honor M. M. Tillet who in 1755 worked with this pathogen in wheat. Since then two species of *Tilletia* have been shown to be involved in this disease *T. tritici* (Syn: *T. caries*) and *T. laevis* (Syn: *T. foetida*). Reticulate, globose and rough walled. No resting period. Germinate to produce primary sporidia which unite to form 'H' shaped structure.

Life cycle

The spores on the seed surface germinate along with the seed. Each produces a short fungal thread terminating in a cluster of elongated cells. These then produce secondary spores which infect the coleoptiles of the young seedlings before the emergence of the first true leaves. The mycelium grows internally within the shoot infecting the developing ear. Affected plants develop apparently normally until the ear emerges when it can be seen that grain sites have been replaced by bunt balls. In India disease occurs only in Northern hills, where wheat is grown.

Favourable conditions

»   Temperature of 18-20°C.
»   High soil moisture.

Disease cycle

Externally seed borne

Management

»   Grow the crop during high temperature period.

»   Adopt shallow sowing.

»   Grow resistant varieties like Kalyan Sona, S 227, PV 18, HD 2021, HD 4513 and HD 4519.

Diseases of Wheat   55

» Seed treatment with skimmed milk powder or skimmed milk 160 g/ kg.

» Seed treatment with carboxin 0.2%, chloronil 2-4 g/ kg or bitertinol 19 µg/ kg.

» Spray with copper sulphate 2% or formalin 0.25% or copper carbonate or propiconazole 0.01%.

9. Karnal bunt

Karnal bunt or partial bunt, caused by *Tilletia indica* (Syn: *Neovossia indica* [Mitra] Mundkur) occurs endemically in India (Mitra, 1931). The first report of a new bunt disease in wheat came from the region of Faizalabad (Pakistan) in 1909.

Karnal bunt was first reported by Mitra (1931) from Karnal, a place now in Haryana state of India from where the name Karnal originates. Karnal bunt is also known as partial bunt. Earlier it was a minor disease found only in Northwestern India. Durum wheat and triticale are less susceptible. Plants are infected within 2-3 weeks of heading.

Symptoms

» Symptoms of Karnal bunt are often difficult to distinguish in the field due to the fact that incidence of infected kernels on a given head is low.

» There may be some spreading of the glumes due to sorus production but it is not as extensive as that observed with common bunt.

» Symptoms are most readily detected on seed after harvest.

» The black sorus, containing dusty spores is evident on part of the seed, commonly occurring along the groove.

» Heavily infected seed is fragile and the pericarp ruptures easily.

» The foul, fishy odor associated with common bunt is also found with karnal bunt.

» The odor is caused by the production of trimethylamine by the fungus.

» Seed that is not extensively infected may germinate and produce healthy plants.

Fungus: *Tilletia indica* Mitra; Syn: *Neovossia indica* (Mitra) Mundk.

Globose and smooth walled. Possess long resting period. Germinate to produce primary sporidia which are needle shaped and then secondary sporidia which is sickle shaped.

Disease cycle

Karnal bunt was reported as a soil borne disease by Mitra (1931), but now it is considered as an air borne disease. The fungal spores are also transferred by means of equipment, tools or by man moving from milling places. The spores remain viable for several years in soil, wheat straw and farm yard manure. Soil or seeds are primary sources of inoculum. Environment plays a key role in disease progression. Teliospores germinate at suitable temperature (15-25°C) and humidity in the soil. This condition generally dominates during February to March in North Indian plains.

On germination, each teliospore produces promycelium which bears 110-185 primary sporidia at its tip. The primary sporidia are sickle shaped and were regarded as infective entities. Now it is well known that secondary sporidia (allantoid and filiform) play an important role in the disease cycle of the pathogen. The allantoid sporidia are pathogenic while filiform sporidia increase the inoculum by division on host/ soil surface. The sporidia are mostly binucleate and on germination produce a germ-tube that penetrates the developing grain through stigma or ovary wall. Infection takes place mainly at the time of anthesis. Generally, the grains are moderately affected but in severe conditions whole grain may be infected.

Favourable conditions

> » Temperature at 15-20°C.

> » High humidity and cloudy weather during flowering.

Development of Karnal bunt depends on the availability of favorable conditions for infection and disease development from heading to flowering (anthesis) of the wheat crop. Progression of disease is favoured by moderate temperatures, high relative humidity or free moisture, cloudiness, and rainfall during anthesis. There is a range of conflicting information available on how abiotic conditions during the rest of the year affect survival of the pathogen and development of Karnal bunt.

Management

> » Deep ploughing during summer season.

> » Avoid continuous cropping of wheat in the same field.

> » Grow resistant varieties like HD 1907, HD 1982 (Janak), HD 2009 (Arjun), HD 2281, HD 2329, DWL 1562, DWL 5023, HD 2283, HD 2285, HI 358, HP 743, L 176, L 191, WL 1562, WL 711, UP 319 and M 137-A.

Diseases of Wheat

» Seed treatment with hexachlorobenzene and cyano (methyl mercuric) guanidine or chloranil 0.2% prevented germination of teliospores of *T. indica*.

» Polyethylene mulching increased the soil surface temperature to 54.5°C and straw burning raised the temperature of the soil surface, 5 cm and 10 cm depth of soil to 92.5°C, 67.0°C and 58.0°C, respectively, and enabled the soil to be disinfected from the bunt propagules.

» Spray propiconazole 0.1%, carbendazim 0.1% and triadimefon 0.1%.

» Fumigation of soil with methyl bromide, metham-sodium and formaldehyde.

10. Powdery mildew

Symptoms

» Greyish white powdery growth appears on the leaf, sheath, stem and floral parts.

» Powdery growth later become black lesion and cause drying of leaves and other parts.

Fungus: *Blumeria graminis* f. sp. *tritici* (DC.) Speer;
  Syn: *Erysiphe graminis var. tritici*

Fungus produces septate, superficial, hyaline mycelium on leaf surface with short conidiophores. The conidia are elliptical, hyaline, single celled, thin walled and produced in chains. Dark globose cleistothecia containing 9-30 asci develop with oblong, hyaline and thinwalled ascospores.

Disease cycle

Fungus remains in infected plant debris as dormant mycelium and asci. Primary spread is by the ascospores and secondary spread through airborne conidia.

Favourable conditions

» Temperature of 20-21°C.

Management

» Grow resistant varieties like Kalyan Sona, Sharbati sonara, C 591, E 750, HD 2189, HD 2278, NP 710, UP 1109 and VL 421.

58   Diseases of Field Crops and their Management

» Spray wettable sulphur 0.2% or triademefon at 500 g/ ha.

» Spray *Ampelomyces quisqualis* mycoparasite 2 lit/ ha.

11. Leaf blight

Symptoms

» Reddish brown oval spots appear on young seedlings with bright yellow margin.

» In severe cases, several spots coalesce to cause drying of leaves.

» It is a complex disease, having association of *A. triticina, B. sorokiniana* and *A. alternata.*

Fungi: *Alternaria triticina* Prasada & Prabhu, (1963);
         Bipolaris sorokiniana (Sacc.) Shoemaker (1959)

Disease cycle

» Primary spread is by externally seed-borne and soil borne conidia.

» Secondary spread is by air-borne conidia.

Favourable conditions

» Temperature of 25°C and high relative humidity.

Management

» Grow resistant varieties like Arnautka, E 6160 and K 7340.

» Soak the seeds in water for 4 hrs followed by 10 min. dip in hot water at 52°C.

» Spray the crop with mancozeb or zineb at 2 kg/ ha.

12. *Pythium* foot rot

Chromista: *Globisporangium abappressorium* (Paulitz & M. Mazzola)
         Pythium graminicolum Subraman., (1928) and *P. arrhenomanes* Drechsler

Diseases of Wheat

Symptoms

- » The disease mainly occurs in seedlings and roots and rootlets become brown in colour.
- » Seedlings become pale green and stunted growth.
- » Parasite produces sporangia, zoospores and oospores.

13. Pink snow mold

Fungus: *Microdochium nivale* (Fr.:Fr.) Samuels & I. C. Hallett
  Syn: *Calonectria nivalis* Schaffnit; *Fusarium nivale* (Fr.) Sorauer

Symptoms

- » Most common early symptoms are dark brown streaks on seedling leaf sheaths as shown above.
- » As the infection progresses, these become water-soaked at stem bases and extend up the first internode.
- » Infections at the ligule then transfer mycelium onto newly emerging leaves, and the fungus is transferred up the growing plant.

14. *Fusarium* foot rot (Dryland foot rot / Strawbreaker foot rot / Ear (Head) blight)

The disease infects wheat, barley, oats, triticale and grasses

Fungi: *Fusarium culmorum* and *F. pseudograminearum* may infect roots and crowns of spring and winter wheat.

This disease is common in dryland winter wheat and no-till annual spring cereals. It is associated with high fall soil temperatures, low fall soil moisture, and moisture stress after anthesis. Oats, barley, and numerous grasses are susceptible.

- » Symptoms in late falls and early spring, roots are brown, and the subcrown internode is discolored.
- » At boot stage, roots and subcrown internode are uniformly dark brown.

60   Diseases of Field Crops and their Management

» The lower stem extending above the first node may be streaked or uniformly brown.

» After heading, white heads may develop that may be void of kernels or produce shriveled kernels.

» Plants may die prematurely.

» Early infections may also cause seed decay or damping-off.

15. Tan spot

Fungus: *Pyrenophora tritici-repentis* (Died.) Drechsler, (1923)
» Tan colored and dimond shaped spots surrounded with yellow halo are formed on the leaf.

» When plant matures, the fungus invades the straw and produce black colored raised fruiting bodies called pseudothecia are formed.

16. Tundu

Other name: Ear cockle, Spike blight, Yellow ear rot, Gumming, Yellow slime

In India, is a complex disease of wheat spikes occurring in association with *Clavibacter tritici* (early) or *Rathayibacter tritici* (Carlson & Vidaver ex Hutchinson) (*Corynebacterium tritici, Corynebacter michiganense* pv. *tritici,*) and the seed gall nematode *Anguina tritici*. *Anguina tritici* was the first plant parasitic nematode to be described in the literature in 1743. It causes a disease in wheat and rye called "ear-cockle" or seed gall. Originally found in many parts of the world but has been eradicated from the western hemisphere. In India it was first reported in Punjab by Hutchinson (1917). The bacterium sticks to the cuticle of the nematode.

Symptoms

» Yellow slime exudation was observed on the stem and inflorescens.

» Slime dries up to form sticky yellow layers and cause curling, twisting and distortion of ear heads and rotting of the spikes.

» Most of the grains in the earhead are replaced by galls formed by the nematodes and galls carry bacterium.

Bacterium: *Clavibacter tritici* (Hutchinson 1917) Burkholder 1948

Diseases of Wheat

Nematode: *Anguina tritici* (Steinbuch, 1799) Filipjev, 1936

Management

- » Grow resistant varieties like Sharbati, Sonora and E 9176.
- » Sieve method (mechanical) sieves of different mesh sizes are used and galls are separated from the seed.
- » Soaking of seeds in brine solution (NaCl or KCl) 20% (14.5 kg/ 450 lit) to separate the galls.
- » Presoaking for 2 hours and then putting grains in hot water at 122°F (50°C) for at least 2 hours. After this grains must immediately be dried by putting them in thin layers.

17. Bacterial leaf streak and black chaff

Symptoms

- » Bacterial streak symptoms can be observed at any stage of plant growth.
- » Initially, small yellowish water-soaked, oily, translucent streaks can be observed on the leaves.
- » Under humid and cold conditions abundant exudation of the bacteria can be observed on the leaves and on the stems as small yellowish granules which later become hard shiny crystals.
- » On the stems, initially, water-soaked yellowish patches can be observed with whitish exudation, becoming dark brown to violet in color.
- » Upon ageing the streaks on leaves coalesce into light brown blotches which later become dark brown and with abundant exudation giving an appearance of whitish-yellow crystals.

Bacterium: *Xanthomonas translucens* (Jones *et al.* 1917) Vauterin *et al.* 1995
Syn: *Xanthomonas translucens* pv. *undulosa* (Smith *et al.*) Vauterin *et al.*

Disease cycle

- » The pathogen is spread by seed, splashing rain, sprinkler and furrow irrigation, plant-to-plant contact, and spike-visiting insects such as aphids.

62   Diseases of Field Crops and their Management

» The pathogen can survive between susceptible small grain crops in plant debris, pathogenically and epiphytically on other crops and weeds, and in seed.

Favourable conditions

» Rainfall and cloudy weather

Management

» Crop rotation

» Controlling volunteer plants and grasses

» Hot-water treatment at 53°C for 10 minutes followed by immediate cooling and drying.

18. Soil borne mosaic disease

» *Soil-borne mosaic virus* encompasses different strains showing variable symptoms between whitish-green to yellow mosaic on leaves.

» The leaves of infected plants are stunted, mottled and show chlorotic stripes parallel to veins.

» Rye, barley and *Bromus* spp. are the hosts of this virus.

» The virus is transmitted through the obligate fungus parasite *Polymixa graminis.*

Minor diseases

» **Downy mildew (Crazy top):** *Sclerophthora macrospora* (Sacc.)

» **Phoma leaf spot:** *Phoma insidiosa* Tassi

» **Pink snow mold:** *Fusarium nivale* (Fr.) Sorauer; *Microdochium nivale* (Fr.:Fr.)

    » **Teleomorph -** *Calonectria nivale* Schaffnit

» **Spot blotch -** *Bipolaris sorokiniana* (Sacc.) Shoemaker

» **Seedling blight and Root rot:** *Rhizoctonia solani* J.G. Kühn AG-8

» **Sclerotinia rot:** *Sclerotinia sclerotiorum* (Lib.) de Bary (1884)

» ***Wheat spindle streak virus***

» ***Wheat streak mosaic virus***

» ***Wheat yellow mosaic virus***

Diseases of Wheat 63

Common diseases on Wheat and Barley

19. Take all disease

Fungus: *Gaeumannomyces graminis* (Sacc.) Arx & D.L. Olivier var. *avenae*.
Syn: *Ophiobolus graminis* var. *avenae* E.M. Turner

» Take-all is most obvious near heading on plants growing in moist soil.

» Diseased crops appear uneven in height and irregular in maturity.

» Diseased plants easily break free at their crown when pulled from the soil.

» Infected plants are stunted, mildly chlorotic, have few tillers and ripen prematurely. Their heads are often bleached (white-heads) and sterile.

» Roots are sparse, blackened and brittle from fungus invasion.

» A black-brown, dry rot extends to the crown and basal stem, where a dark skinny fungus plate is diagnostic just beneath the lowest leaf path.

20. Scab (The *Fusarium* Head Blight)

Fungus: *Fusarium graminearum* Schwabe *sensu lato*

» Infected plants shows premature death or blighting of the spikelets.

» Bleached spikelets usually are sterile or contain only partially filled seeds.

21. Leaf blotch

» *Mycosphaerella graminicola* (Fuckel) J.Schröt (*Septoria tritici, S. avenae* Desm.)

» *Leptosphaeria nodorum* E. Müll. (*Stagnospora nodorum*)

» *Leptosphaeria avenaria* G. F. Weber (*Parastagonospora avenae* A.B.Frank)

22. Molya disease

Nematode: *Heterodera avenae* Wollenweber, 1923

» Cereal cyst nematode or Ustinov cyst nematode infection on wheat and barley.

It is also known as cereal cyst, it prevalent in isolated pockets in Rajasthan, Haryana,

Punjab and Himachal Pradesh. Its incidence is higher in loamy and sandy soils as compared to medium and heavy soils.

Symptoms

- » The symptoms of disease include stunting of plants, discoloration of leaves of resembling nutrient deficiency and poor tillering.
- » The infected plants are stunted, pale and unhealthy.
- » Under heavy infection, the plant may not develop beyond 15 cm in height.
- » Roots of infested plants are stunted with along main root and few rootlets at the extreme end.

Management

- » Adoption of cultural practices can control the disease.
- » Ploughing the fields two or three times in the month of May-June and are rotating the wheat crop with non-host crops such as gram, carrot, radish, marigold and resistant varieties of barley for one and two years.
- » Mixing practices i.e. carbofuran at 1.5 kg/ha in the soil also controls the disease to greater extent.

Minor diseases

- » **Basal glume rot and bacterial leaf blight**:

  Pseudomonas syringae pv. *atrofaciens* (McCulloch) Young *et al*

- » *Barley stripe mosaic virus*
- » *Barley yellow dwarf virus*
- » *Barley yellow mosaic virus*
- » *Barley mild mosaic virus*

Chapter - 3

Diseases of Barley - *Hordeum vulgare* L.

S. No.	Disease	Pathogen
1.	Black (stem) rust	*Puccinia graminis* f. sp. *tritici*
2.	Brown (leaf) rust	*Puccinia hordei*
3.	Yellow (stripe) rust	*Puccinia striiformis* f. sp. *hordei*
4.	Loose smut	*Ustilago tritici* (*U. nuda*)
5.	Covered smut	*Ustilago hordei*
6.	Spot blotch	*Bipolaris sorokiniana*
7.	Leaf scald	*Rhynchosporium secalis*
8.	Net type net blotch	*Pyrenophora teres* f. sp. *teres*
9.	Powdery mildew	*Erysiphe graminis* f. sp. *hordei*
10.	Barley yellow dwarf disease	*Barley yellow dwarf virus*

Barley rusts

Rust fungi infect and damage crop plants but those infecting the cereal crops such as wheat, barley, rye, and oats historically caused the biggest commercial losses and pose the largest threats to world food security.

Diseases of Field Crops and their Management

1. **Black (stem) rust:** *Puccinia graminis* Pers.:Pers f. sp. *tritici* Eriks. & E. Henn.

   » Dark reddish brown pustules on both sides of leaves, stems and on spikes are generally appeared.

   » It requires warmer temperature for infection

2. **Brown (leaf) rust:** *Puccinia hordei* Otth

   » The symptom appears randomly on upper surface of leaf and leaf sheaths and occasionally on neck and awns as small orange/ orange brown pustules, primarily.

   » The temperature requirement is between 20-25°C.

   » Pycnial and aecial stage found in Star of Bethlehem (*Ornithogalum umbellatum*)

3. **Yellow (stripe) rust:** *Puccinia striiformis* Westend. f. sp. *hordei* Eriks. & Henn.

   » The narrow stripes containing yellow to orange yellow colour pustules on leaf sheaths, necks and glumes appears in the stripe rust.

   » Yellow rust is the disease of cool temperature (10-20°C) and the availability of free moisture is further congenial to spread infection.

4. **Crown rust:** *Puccinia coronata* Corda. f. sp. *hordei* Jin & Steff.

   » Uredinia are linear, light orange, and occur mostly on the leaf blades but occasionally occur also on leaf sheaths, peduncles and awns.

   » Extensive chlorosis is often associated with the uredinia.

   » Telia are mostly linear, black to dark brown, and are covered by the host epidermis.

   » Alternate host is common buckthorn (*Rhamnus cathartica*)

Management

   » Crop rotation with non-host crops and removal of infected crop debris.

   » Grow multiple disease resistant cultivars like RD 2508, RD 2035, DWRUB 52, RD 2552 and RD 2624.

   » Grow yellow rust resistant cultivars like DWRUB 52, DWRB 73, DWRUB 64, DWRB 91 and DWRB 92.

Diseases of Barley

» Seed treatment with carboxin 2-3 g/ kg or tebuconazole 1 g/ kg.

» Spray azoxystrobin 25% EC 0.1% or triadimefon 25% EC 0.1% or difenoconazole 25% EC 0.1% or propiconazole 25% EC 0.1% or tebuconazole 25% EC at 0.1%.

Barley smuts

Loose smut (*Ustilago avenae*) and covered smut (*Ustilago hordei*) of barley and oats are externally seed-borne diseases with similar symptoms which are difficult to distinguish in the field. Both diseases are managed in the same way. After sowing, spores on the seed surface germinate and infect the emerging seedling. The fungus grows without symptoms within the plant and identification of infected plants is difficult prior to head emergence. Affected plants may be slightly taller and heads emerge earlier than the main part of the crop. Each spikelet, including the chaff, is transformed into a spore mass which is at first covered with a fine membrane. This membrane soon bursts releasing the spores to contaminate healthy heads, leaving a bare stalk or rachis on the infected plant.

5. **Loose smut:** *Ustilago tritici* (Pers.) Rostr. *Ustilago nuda* (C.N. Jensen) Rostr., nom. nud.

» The entire inflorescence gets turned to smutted head containing black powdery masses.

» The disease is caused by the internally seed borne pathogen and expresses only at the time of flowering.

» The losses in infected spikes are 100%.

» The fungus is internally seed borne by dormant mycelium within the embryo of the barley seed.

Management

» Sow healthy seeds and treatment with carboxin 2 g/ kg or carbendazim 2.5 g/ kg.

6. **Covered smut:** *Ustilago hordei* (Pers.) Lagerh

» Masses of dark brown smut spores replace the entire head of plants and spores are contained in a membrane until plant maturity.

» When spores are dislodged by threshing they infect the seed.

68   Diseases of Field Crops and their Management

> The hard spore balls of covered smut are very common in soils of untreated plots harvest.

> Spread by externally seed borne and systemic in natire

> During the threshing process, teliospores are released to contaminate the healthy barley seeds

Management

> Grow resistant varieties like BHs 4, BM 23, LB 873, EB 4003, K 24 and NP 109.

> Seed treatment with carboxin and thiram (1:1) or tebuconazole at 1.5 g/ kg or agrosan (phenyl mercury acetate) or ceresin wet or carboxin 2.5 g/ kg or carbendazim 2.0 g/ kg myclobutanil 0.5 g/ kg.

7.   Spot blotch

Fungus: Anamorph: *Bipolaris sorokiniana* (Sacc.) Shoemaker
          Teleomorph: *Cochliobolus sativus* (Ito & Kuribayashi) Drechs. ex Dastur

a. Leaf blight

> The disease spreads as small light brown spindle spots distributed on leaf blade increasing in size along the leaf veins.

> The spots are irregular and vary from oval to oblong or elliptical.

> Fully developed lesions become dark brown colour and cover entire leaf by merging together.

b. Seedling blight, kernal blight (Black point), foot rot and root rot

> The causal fungus affects all parts of the plant and produces a variety of symptoms, from a seedling blight and root rot to "Black point" of the kernels.

> Brown spot initially appear as chocolate brown-to-black appear near the soil line or at the base of the sheaths that cover the seedling leaves and on the root, internodes spread over entire seedlings causing blight.

> Spot with straw colour centre and brown margin produced on the leaves. Infected plant produces fewer tiller, few seed, ripen prematurely, kernel in the ear head dark brown and shrivelled.

Diseases of Barley 69

Management

» Grow spot and net blotch disease resistant cultivars like DWRUB 52, DWRB 73, DWRUB 64 and RD 2552.

» Sow healthy seeds.

» Spray with triadimefon 25% EC 0.1% or propiconazole 25% EC 0.1% or tebuconazole 25% EC at 0.1%.

8. Leaf scald

Fungus: *Rhynchosporium secalis* (Oudem.) J.J.Davis

» The disease causes scald-like lesions of the leaf blades and sheaths.

» The spots are oval shaped to irregular with straw-colored centers and brown margin.

» In severe infections, the lesion coalesces and cause defoliation

9. Net type net blotch (NTNB)

Fungus: Anamorph: *Drechslera teres* (Sacc.) Shoemaker
Teleomorph: *Pyrenophora teres* f. sp. teres Drechs.

» It appears as small circular brown spots that develop into a chocolate brown net-like pattern on leaves, leaf sheaths and glumes with yellowing of the areas surrounding the net pattern.

10. Powdery mildew

Fungus: Teleomorph: *Erysiphe graminis* DC. f. sp. *hordei* Em. Marchal
Syn:*Blumeria graminis* (DC.) E.O. Speer
Anamorph: *Oidium monilioides* (Nees) Link

» The symptoms appear as small white grey dots on leaf surface and change to grey or greyish brown at later stages.

» In case of susceptible varieties, the spots coalesce and form large necrotic blotches and in severe cases the leaves dry prematurely.

» The disease is favoured by cool, cloudy and humid weather conditions.

11. *Barley yellow dwarf virus (Luteovirus)*

» Symptoms caused by BYDV are ambiguous and often overlooked or associated with nutritional or non-parasitic disorders.

» BYD is diagnosed by the presence of aphid vectors and occurrence of yellowed, stunted plants, singly or in small groups.

» Seedling infections slow plant growth and cause prominent or yellowing of old leaves.

» Yellowed or reddened flag leaves on otherwise normal plants are indicative of pest seedling infections.

» Diseased plants have less flexible leaves and underdeveloped root systems.

» Phloem tissues may be darkened.

» Cool temperatures (16-20°C) enhance symptoms expression.

Vector: *Rhopalosiphum padi, Rhopalosiphum maidis* (Aphids).

Minor diseases of barley

» **Downy mildew (Crazy top):** *Sclerophthora rayssiae*

» ***Pythium* root rot**: *Pythium arrhenomanes, Pythium graminicola,*

» Pythium tardicrescens

» ***Rhizoctonia* root rot**: *Rhizoctonia solani, Rhizoctonia oryzae*

» ***Stagonospora* blotch**: *Stagonospora avenae* f. sp. *triticea* T. Johnson

Phaeosphaeria avenaria f. sp. *triticea* T. Johnson

» **Sclerotial foot rot:** *Pellicularia rolfsii* (Curzi) E. West

» **Leaf stripe:** *Pyrenophora graminea* Ito & Kuribayashi [Anamorph: *Drechslera graminea* (Rabenh.) Shoemaker] or *Helminthosporium gramineum* f. sp. *tritici-repentis* Died.

» **Spot type net blotch** (STNB): *Pyrenophora teres* Drechs. f. sp. *maculate*

» ***Septoria* speckled leaf blotch** (SSLB): *Septoria passerinii*

» **Root rot or Foot rot**: *Fusarium culmorum* (W.G. Smith) Sacc.

Diseases of Barley

» **Glume blotch**: *Parastagonospora nodorum* (Berk.) Quaedvlieg, Verkley & Crous

» ***Leptosphaeria* leaf spot:** *Leptosphaeria herpotrichoides* De Notaris

» **Ergot:** *Claviceps purpurea* (Fr.) Tul. or *Sphacelia segetum* Lév.

» **Black chaff and bacterial streak:** *Xanthomonas translucens* pv. *translucens* (ex Jones *et al.*, 1917) Vauterin *et al.* 1995

» **Bacterial leaf and kernel blight**: *Pseudomonas syringae* pv. *syringae* van Hall 1902

» **Bacterial stripe**: *Pseudomonas syringae* pv. *striafaciens* (Elliott 1927) Young *et al.* 1978

» **Basal glume rot**: *P. syringae* pv. *atrofaciens* (McCulloch 1920) Young *et al.* 1978

» ***Barley mosaic virus*:** Transmitted by *Rhapalosiphum maidis*

» ***Barley stripe mosaic virus*** (BSMV) genus *Hordeivirus*

» ***Barley yellow mosaic virus*** (BaYMV) genus *Bymovirus*

» ***Barley yellow streak mosaic virus*** (BYSMV)

Chapter - 4

Diseases of Oats - *Avena sativa* L.

S. No.	Disease	Pathogen
1.	Leaf blotch	*Septoria avenae* f. sp. *avenae*
2.	Leaf scald	*Rhynchosporium secalis*
3.	Crown rust	*Puccinia coronata*

1. Leaf blotch

Symptoms

- » This disease appears as reddish tan leaf blotches that are somewhat linear with irregular margins.

- » Heavily infected leaves die.

- » Seedling blight may occur if coleoptiles are infected.

Fungus: Anamorph: *Septoria avenae* f. sp. *avenae* Johnson

Teleomorph: *Phaeosphaeria avenaria* f. sp. *avenaria* Shoemaker & C.E. Babc

Fungus establishes itself inter-cellular, tends to produce conidiophores which emerges through stomata and bears single terminal conidia. These conidia are brown to black in color, tapered at apex, 4-6 spectate.

Diseases of Oat

» Primarily infection occurs through these conidia.

» Secondary infection occurs through air-borne spores.

Management

» Seed treatment with carbendazim 0.1% or captan 2.5%.

» Spray with carbendazim + mancozeb 0.1% or chlorothalanil 0.1% or zineb 2.5%.

2. Leaf scald of barley and oats

Fungus: *Rhynchosporium secalis* (Oudem.) J.J.Davis

» Symptoms first appear as chlorotic, irregular or diamond-shaped lesions.

» Later symptoms are typically blue-grey water-soaked lesions on leaves and leaf sheaths.

» Mature lesions become pale brown with a dark purple margin.

» As they grow they merge forming large areas of dead tissue, even destroying the whole plant leaf area.

3. Crown rust

Fungi: *Puccinia coronata* Corda; *Puccinia coronata* Cda f. sp. *avenae* Erikss.

» The characteristic symptom is the development of round to oblong, orange to yellow pustules, primarily on leaves but also on stems and heads.

» The powdery spore masses in the pustules are readily dislodged.

» The pustule areas turn black with age.

» Losses result from damage to leaves (particularly the flag leaf), which leads to reduced photosynthesis and transport of carbohydrates to the developing grain.

» This causes shriveled grain and reduced grain quality.

» Alternate host is Buckthorn (*Rhamnus* spp).

Minor diseases

- » **Root rot**: *Fusarium* spp.; *Pythium* spp.; *Pythium debaryanum* Auct. non R. Hesse *Pythium irregulare* Buisman; *Pythium ultimum* Trow

- » **Stem rust**: *Puccinia graminis* Pers. f. sp *avenae* Erikss. & Henn.

- » **Loose smut**: *Ustilago avenae* (Pers.) Rostr.

- » **Covered smut**: *Ustilago segetum* (Bull.:Pers.) Roussel (*Ustilago kolleri* Wille)

- » **Powdery mildew:** *Erysiphe graminis* DC. f. sp. *avenae* Em. Marchal

- » **Downy mildew:** *Sclerophthora macrospora* (Sacc.) Thirumalachar *et al.*

- » ***Pyrenophora* leaf blotch**: *Pyrenophora avenae* Ito & Kuribayashi

- » **Victoria blight**: *Bipolaris victoriae* (F. Meehan & Murphy) Shoemaker

    Teleomorph: *Cochliobolus victoriae* R.R. Nelson

- » **Seed and seedling blight**: *Rhizoctonia* sp., *Fusarium* sp., and *Helminthosporium* sp.

- » **Red leather leaf:** *Spermospora avenae* Sprague & A.G. Johnson

- » **Bacterial blight (halo blight):** *Pseudomonas coronafaciens* pv. *coronafaciens* (Ell.)

- » **Bacterial stripe blight**: *Pseudomonas coronafaciens* pv. *striafaciens* (Elliott) Young.

- » **Black chaff and bacterial streak:** *Xanthomonas translucens* pv. *translucens* (ex Jones *et al.*, 1917) Vauterin *et al.* 1995

- » ***Oat mosaic virus***

Chapter - 5

Diseases of Maize/ Corn - *Zea mays* L.

S. No.	Disease	Pathogen
1.	Sorghum downy mildew	*Peronosclerospora sorghi*
2.	Brown stripe downy mildew	*Sclerophthora rayssiae* var. *zeae*
3.	Philippine downy mildew	*Peronosclerospora philippinensis*
4.	Sugarcane downy mildew	*Peronosclerospora sacchari*
5.	Crazy top	*Sclerophthora macrospora*
6.	Java downy mildew	*Peronosclerospora maidis*
7.	Green ear	*Sclerospora graminicola*
8.	Rajasthan downy mildew	*Peronosclerospora heteropogoni*
9.	Leaf-splitting downy mildew	*Peronosclerospora miscanthi*
10.	Spontaneous downy mildew	*Peronosclerospora spontanea*
11.	Turcicum leaf blight	*Exserohilum turcicum*
12.	Maydis leaf blight	*Bipolaris maydis*
13.	Banded leaf and sheath blight	*Rhizoctonia solani* f. sp. *sasakii*
14.	Tropical rust	*Phakopsora zeae*
15.	Common rust	*Puccinia sorghi*
16.	Southern rust	*Puccinia polysora*

17.	Common smut	*Ustilago maydis*
18.	Head smut	*Sporisorium reilianum*
19.	False head smut	*Ustilaginoidea virens*
20.	Charcoal post flowering stalk rot	*Macrophomina phaseolina*
21.	*Pythium* stalk rot	*Pythium aphanidermatum*
22.	*Fusarium* stalk rot and wilt	*Fusarium* spp.
23.	Brown spot	*Physoderma maydis*
24.	*Curvularia* leaf spot	*Curvularia aeria*
25.	Grey leaf spot	*Cercospora zeae-maydis*
26.	Anthracnose	*Colletotrichum graminicola*
27.	*Rostratum* leaf spot	*Exserohilum rostratum*
28.	Bacterial stalk rot	*Erwinia dissolvens*
29.	Maize dwarf mosaic	*Maize dwarf mosaic virus*
30.	Maize mosaic	*Maize mosaic virus I*
31.	Maize streak	*Maize streak virus*

Downy mildew

Downy mildews are important maize diseases in many tropical regions of the world. They are particularly destructive in many regions of tropical Asia where losses in excess of 70% have been documented. Downy mildews are caused by up to ten different species of oomycete (Chromista) pathogen includes the genera like *Peronosclerospora*, *Scleropthora* and *Sclerospora*. Downy mildews originated in the Old World although they have since been introduced to many regions of the New World.

Symptoms of downy mildew on maize caused by the various pathogenic species are similar, although symptoms can vary depending on plant age, prevailing climatic conditions, and host germplasm. Infection of maize plants at the seedling stage (less than 4 weeks old) results in stunted and chlorotic plants and premature plant death.

Common symptom of maize downy mildew

> » The most characteristic symptom is the development of chlorotic streaks on the leaves.

Diseases of Maize 77

» Dull white or grey-white downy growth is seen on the lower surface of leaf.

» Downy growth also occurs on bracts of green unopened male flowers in the tassel.

» Plants exhibit a stunted and bushy appearance due to shortening of the internodes.

» Small to large leaves are noticed in the tassel.

1. Sorghum downy mildew

Chromista: *Peronosclerospora sorghi* (W.Weston & Uppal) C.G. Shaw;
Syn: *Sclerospora sorghi* W. Weston & Uppal

Symptoms

» Leaves on older plants display characteristic symptoms of downy mildews which include mottling, chlorotic streaking and lesions, and white striped leaves that eventually shred.

» Downy growth is often observed on both leaf surfaces, but is more common on the lower leaf surface.

» Infected plants have leaves that are narrower and more erect compared to healthy leaves. Infected plants are often stunted, tiller excessively and have malformed reproductive organs (tassels and ears).

» Infected plants may not seed, while tassels may exhibit 'bushy' growth.

» The disease has been known to occur through a collateral host *Heteropogen centortus* on which the fungus perpetuates of the host.

2. Brown stripe downy mildew

Chromista: *Sclerophthora rayssiae* Kenneth & I. Wahl var. *zeae* Payak & Renfro

Symptoms

» Most destructive downy mildew pathogen of maize crop.

» Lesions start developing on lower leaves as narrow chlorosis or yellow stripes, 3-7 mm wide, with well defined margin and are delimited by the veins.

» The stripes later become reddish to purple.

» Lateral development of lesions causes severe striping and blotching.

» Seed development may be suppressed.

» Plant may die prematurely if blotching occurs prior to flowering.

» Sporangia on the leaves appear as a downy whitish to wooly growth on both surface of the lesions.

» Floral or vegetative parts are not malformed, and the leaves do not shred.

3. Philippine downy mildew

Chromista: *Peronosclerospora philippinensis* (W.Weston) C.G.Shaw

Symptoms

» The symptoms are very similar to that of sugarcane downy mildew except for intensity of color of stripes.

» The leaf infection results in long, chlorotic stripes with downy fungal growth.

» Affected plants produced malformed tassels or aborted ears which may appear any time to silking, but the plants affected early are stunted and often die.

4. Sugarcane downy mildew

Chromista: *Peronosclerospora sacchari* (T.Miyake) C.G.Shaw
        Syn: *Sclerospora sacchari* T. Miyake

Symptoms

» Development of long, rather broad chlorotic stripe along almost entire length of leaves.

» It may be whitish in earlier stage changes to dark brown yellowin very late stages.

» The stripes appear on young leaves following infection of the growing point.

» The fungal growth seen on both the surfaces of the stripes.

» Typically, yellow to white stripes, and the white downy growth that occurs on both sides of the leaves.

Diseases of Maize

» The affected plants may be malformed with undeveloped tassels and ears. The ears are poorly filled.

5. Crazy top

Chromista: *Sclerophthora macrospora* (Sacc.) Thirumalachar *et al.*
Syn: *Sclerospora macrospora* Sacc.

Symptoms

» Proliferation of auxillary buds on the stalk of tassel and the cobs is common and it is called as Crazy top.

6. Java downy mildew

Chromista: *Peronosclerospora maidis* (Racib.) C.G.Shaw

» Infection of maize plants at the seedling stage results in stunted and chlorotic plants and premature plant death.

» Leaves on older plants display characteristic symptoms of downy mildews which include mottling, chlorotic streaking and lesions, and white striped leaves that eventually shred

7. Green ear

Chromista: *Sclerospora graminicola* W.Weston & Uppal

» Entire ear (inflorescence) is transformed into green leafy mass.

» The lower part of the inflorescence is converted to green leafy mass but the upper part bears seeds.

» Incidence in maize is rarely reported.

8. Rajasthan downy mildew

Chromista: *Peronosclerospora heteropogoni* Siradhana, Dange, Rathore & S.D. Singh

» In seedling a complete chlorosis or chlorotic strips is appeared.

80   Diseases of Field Crops and their Management

» The symptoms are are chlorotic, leaves of infected plants tend to be narrower and more erect than these healthy plants appear at two to three leaf stages.

» Tassels may be malformed producing less pollen while ears may be aborted resulting partial or complete sterility.

» In early symptoms plants are stunted and may die

9.   Leaf-splitting downy mildew

Chromista: *Peronosclerospora miscanthi* (T.Miyake) C.G.Shaw;
Syn: *Sclerospora miscanthi* T. Miyake

» Maize is susceptible to *Peronosclerospora miscanthi* when inoculated, but infection has not been found to occur naturally.

10.   Spontaneous downy mildew

Chromista: *Peronosclerospora spontanea* (W.Weston) C.G.Shaw
Syn: *Sclerospora spontanea* W. Weston

» On young leaves, pale yellow to white streaks or stripes of various lengths appear, extending from the leaf bases to the leaf tips.

» Older leaves show continuous or interrupted yellowish streaks along the blades, the streaks elongate, coalesce with adjacent streaks and turn reddish-brown.

» Leaf twisting and yellowish streaks are observed on some plants.

» Conidia may be produced in the streaks under suitable conditions.

» The shredded tissue contains numerous oospores, which scatter readily in the wind.

Confirmation

The causal agent of downy mildew can be determined by detailed examination of spore-bearing structures (conidiophores or sporangiophores) and spores (conidia or sporangia). Geographic location can be used as a preliminary diagnostic. White stripe symptoms caused by downy mildew are not limited by veins and can therefore be distinguished from signs of iron deficiency.

Diseases of Maize

Chromista

The parasite grows as white downy growth on both surfaces of the leaves, consisting of sporangiophores and sporangia. Sporangiophores are quite short and stout, branch profusely into series of pointed sterigmata which bear hyaline, oblong or ovoid sporangia (conidia). Sporangia germinate directly and infect the plants. In advanced stages, oospores are formed which are spherical, thick walled and deep brown.

Pathogen characters

Chromista (Disease name)	Sporangiophores	Sporangia	Oospores	Initial source of inoculum	Seed borne
Peronosclerospora sorghi (Sorghum downy mildew)	Erect, dichotomously branched, 180 to 300μm in length. Emerge singly or in groups from stomata.	Oval (14.4-27.3 ×15-28.9μm), borne on sterigmata (about 13μm long.	Spherical (36μm in diameter on average), light yellow or brown in color.	Oospores and sporangia	Yes
P. maydis (Java downy mildew)	Clustered conidiophores (150 to 550μm in length) emerge from stomata. Dichotomously branched two to four times.	Spherical to subspherical in shape (17-23μm ×27-39μm).	Not reported.	Sporangia	Yes
P. philippinensis (Philippine downy mildew)	Erect and dichotomously branched two to four times. 150 to 400μm in length and emerge from stomata.	Ovoid to cyclindrincal (17-21μm×27-38μm), slightly rounded at apex.	Rare, spherical (25 to 27μm in diameter and smooth walled.	Sporangia	Yes
P. sacchari (Sugarcane downy mildew)	160 to 170μm in length erect and arise singly or in pairs from stomata.	Elliptical, oblong (15-23μm ×25-41μm) with round apex.	40 to 50μm in diameter, globular, yellow.	Sporangia	

Sclerospora graminicola (Graminicola downy mildew or green ear)	Average length of 268μm.	Borne on short sterigmata, elliptical (12-21×14-31μm) with distinctive papillate operculum at apex.	Pale brown and 22 to 35μm in diameter.	Oospores and sporangia	No
Sclerophthora macrospora (Crazy top)	Very short (14μm on average).	Lemon shaped (30-65×60-100μm), operculate.	Pale yellow, circular (45-75μm).	Oospores and sporangia	No
Scleropthora rayssiae var. *zeae* (Brown stripe downy mildew)		Oval to cyclindrical (18-26× 29-67μm).	Spherical (29-37μm in diameter), brown in color.	Oospores and sporangia	Yes

Disease cycle

The primary source of infection is through oospores in soil and also dormant mycelium present in the infected maize seeds. Secondary spread is through airborne conidia. Depending on the pathogen species, the initial source of disease inoculum can be oospores that over winter in the soil or conidia produced in infected, over wintering crop debris and infected neighboring plants. Some species that cause downy mildew can also be seed borne, although this is largely restricted to seed that is fresh and has high moisture content.

At the onset of the growing season, at soil temperatures above 20°C, oospores in the soil germinate in response to root exudates from susceptible maize seedlings. The germ tube infects the underground sections of maize plants leading to characteristic symptoms of systemic infection including extensive chlorosis and stunted growth. If the pathogen is seed borne, whole plants show symptoms. Oospores are reported to survive in nature for up to 10 years. Once the fungus has colonised host tissue, sporangiophores (conidiophores) emerge from stomata and produce sporangia (conidia) which are wind and rain splash disseminated and initiate secondary infections. Sporangia are always produced in the night. They are fragile and can not be disseminated more than a few hundred meters and do not remain viable for more than a few hours.

Diseases of Maize

Germination of sporangia is dependent on the availability of free water on the leaf surface. Initial symptoms of disease (chlorotic specks and streaks that elongate parallel to veins) occur in 3 days. Conidia are produced profusely during the growing season. As the crop approaches senescence, oospores are produced in large numbers.

Favourable conditions

- » Low temperature (21-33°C)

- » High relative humidity (90%) and drizzling.

- » Young plants are highly susceptible.

Management

- » Deep ploughing and crop rotation with pulses.

- » Collect the debris of the crop and destroy by burning, burying or remove and use as livestockfeed.

- » The eradication of collateral and wild hosts near maize field and rouging infected maize plants has been recommended.

- » Control weeds to increase aeration within the crop and reduce moisture levels in the soil.

- » Ensure seed has low moisture content (below 9%) before planting.

- » Do not rotate or simultaneously cultivate maize with alternate hosts of downy mildew.

- » Do not interplant maize and sugarcane.

- » Do not plant maize close to sugarcane in case of cross-contamination.

- » Plant when soil temp is below 20°C which is unfavorable for oospore germination.

- » Reduce crop density to increase aeration.

- » Grow resistant varieties and hybrids *viz.* Co 1, CoH 1, CoH 2, DMR 1, DMR 5, Amber, Kissan, Ganga safed 2, Ganga 51 and Ganga 11.

- » In addition to above downy mildew resistant varieties PAU 352, Pratap Makka 3, Gujarat Makka 4, Shalimar KG 1, Shalimar KG 2, PEMH 5, Bio 9636, NECH- X 1280 are resistant towards brown stripe downy mildew and DMH 1, NAC6002, COH (M) 4, COH (M) 5, Nithyashree for sorghum downy mildew.

84 Diseases of Field Crops and their Management

» Seed treatment with metalaxyl at 6 g/ kg.

» Spray the crop with metalaxyl + mancozeb at 2.5 g/ lit on 20th day after sowing.

11. Turcicum leaf blight / Northern corn leaf blight

Symptoms

» The fungus affects the crop at young stage.

» Symptoms of turcicum leaf blight are easily recognized.

» Early symptoms are small, oval, water-soaked spots on leaves.

» The spots gradually increase in area into bigger elliptical spots and are straw to grayish brown in the centre with dark brown margins.

» Mature symptoms of turcicum leaf blight are characteristic cigar shaped lesions that are 3 to 15 cm long.

» Lesions are elliptical and tan in color, developing distinct dark areas as they mature that are associated with fungal sporulation.

» Lesions typically first appear on lower leaves, spreading to upper leaves and the ear sheaths as the crop matures.

» Under severe infection, lesions may coalesce, blighting the entire leaf.

» The surface is covered with olive green velvetty masses of conidia and conidiophores.

Fungus: *Exserohilum turcicum* (Pass.) K.J. Leonard & Suggs.
Syn: *Bipolaris turcica* (Pass.) Shoemaker;
Drechslera turcica (Pass.) Subram. & P.C. Jain;
Helminthosporium turcicum Pass.;
Setosphaeria turcica (Luttr.) K.J. Leonard & Suggs;
Trichometasphaeria turcica Luttr.

Conidiophores are in group, geniculate, mid dark brown, pale near the apex and smooth. Conidia are distinctly curved, fusiform, pale to mid dark golden brown with 5-11 septa.

Disease cycle

» It is a seed-borne fungus.

Diseases of Maize

» It also infects sorghum, wheat, barely, oats, sugarcane and spores of the fungus are also found to associate with seeds of green gram, black gram, cowpea, varagu, Sudan grass, Johnson grass and Teosinte.

Favourable conditions

» Optimum temperature for the germination of conidia is 8-27°C provided with freewater on the leaf.

» Infection takes place early in the wet season.

Management

» Grow resistant varieties like CVS 5, Deccan 105, Deccan 109, Sartaj, Trishulata, EH 40146 and JH 1267.

» Seed treatment with captan at 4 g/kg.

» Spray mancozeb 2 kg or captan 1 kg/ha.

» Spray with mancozeb at 2.0 g/ lit or zineb 2.5 g/ lit.

12. Maydis leaf blight / Southern corn leaf blight

Symptoms

» Young lesions are small and diamond shaped, as they mature, they elongate.

» Lesions may coalesce, producing a complete "burning" of large areas of the leaves.

» They vary in size and shape among inbreds and hybrids with different genetic background.

» Symptoms of maydis leaf blight caused by Race T are oval and slightly larger (6-12 × 6-27 mm) than those caused by Race O.

» Lesion borders are usually characterized by dark, brown borders.

» Race T causes lesions on all above ground parts of the plant (including stems, sheaths and ears) and can also cause ear rots.

» Seedlings from seeds infected with Race T often wilt and die within 3 to 4 weeks.

» Under severe disease pressure, usually when infection occurs prior to silking, lesions may coalesce, blighting the entire leaf.

» In these circumstances, sugars may be diverted from the stalk for grain filling, thus predisposing the plant to lodging.

Fungus: *Bipolaris maydis* (Y. Nisik. & C. Miyake) Shoemaker
        Syn: *Cochliobolus heterostrophus* (Drechsler) Drechsler;
        Drechslera maydis (Y. Nisik. & C. Miyake) Subram. & P.C.Jain;
        Helminthosporium maydis Y.Nisik. C.Miyake (Race O, Race C and Race T);
        Ophiobolus heterostrophus Drechsler

The fungus produces long, curved conidiophores through the stomata. The conidia are curved without protruding hilum. Race T is infectious to maize plants with the Texas male sterile cytoplasm (cms-T cytoplasm maize) in USA, leads to epidemic form during 1969-1970.

Disease cycle

» It is a seed-borne fungus.

» It also infects sorghum, wheat, barely, oats, sugarcane and spores of the fungus are also found to associate with seeds of green gram, black gram, cowpea, varagu, Sudan grass, Johnson grass and Teosinte.

Favourable conditions

» Optimum temperature for the germination of conidia is 8 to 27°C provided with freewater on the leaf.

» Infection takes place early in the wet season.

Management

» Grow resistant varieties like Jawahar, Pusa prakash, Ganga 5, Ranjit, Deccan, VL 42, Deccan 109, Prabhat, KH 5901, PRO 324, PRO 339, ICI 701, F 7013, F 7012, PEMH 1, PEMH 2, PEMH 3, Paras and Sartaj.

» Rotation with non-host crops and reduced maize monoculture.

» Spray with mancozeb at 2 g/ lit of water.

Diseases of Maize

13.   Banded leaf and sheath blight

In India, the disease was first reported in 1960 from Tarai region of Uttar Pradesh. The causal organism was *Hypochonus sasakii* which appeared in epidemic form in 1972 in Mandi district of Himachal Pradesh.

Symptoms

» Symptoms of BLSB are conspicuous and characterized with presence of peculiar bands on leaf sheath and sclerotial bodies on affected parts of the plants.

» Symptoms appear on all aerial plant parts except tassel.

» The disease manifests itself on leaf, leaf sheath, stalks and ears as leaf and sheath blight, stalk lesions or rind spotting and stalks breakage, clumping and cracking of styles (silk fibers) and horseshoe shaped lesions with banding of caryopses resulting in ear rots.

» The symptoms appeared within 4-5 days after inoculation, which are irregular, water-soaked, straw-coloured lesions on leaf bases and sheaths.

» The lesions enlarge rapidly resulting in discolored areas alternating with dark bands, apparent on lower leaves after 7-8 days.

Fungus: *Rhizoctonia solani* f. sp. *sasakii* AG-1 Ahuja & Payak (Mycelial & sclerotial stage) *Thanatephorus cucumeris* (Frank) Donk (Basidial stage).

It produces pale to brown colour of mycelium, branching near the distal septum in young growing hyphae, presence of a constriction and formation of a septum in branch near the point of origin, absence of clamp connections, sclerotia of undifferentiated texture, young multinucleate hyphal cells with a prominent septal pore apparatus and rapid growth rate.

Disease cycle

» The primary source is sclerotia in soil or in infected host debris and the active mycelium on the other grass hosts

» Secondary spread is due to contact of healthy plants with infected leaves/sheaths.

Favourable conditions

» Relative humidity of 100% and temperature 30-32°C.

Management

» Remove weed hosts and plant debris before sowing.

» Select well-drained field and planting on raised beds.

» Maintaining the proper population and application of FYM prior to planting.

» Grow resistant hybrids such as EH 1389 and JH 10704

» Seed treatment of peat based formulation of *Pseudomonas fluorescens* at 16 g/ kg or as soil application at 7 g/ lit of water.

» Seed treatment with carbendazim at 2.5 g/ kg or carboxin 2 g/ kg or thiram at 2.5 g/ kg.

» Inter-cropping of maize with legumes especially with soybean.

» Stripping of lower 2-3 leaves along with their sheath considerably lowers incidence and also does not affect grain yield.

» Soil application of *P. fluorescens* and *T. viride* 25 kg/ ha.

» Spray validamycin 0.25% (Sheethmar 27 ml/ lit) followed by *Trichoderma harzianum*

» Spray carbendazim 0.1% or propiconazole 0.15% or copper oxychloride 3 g/ lit other fungicides chloroneb, captafol, mancozeb, zineb, edifenphos, iprobenphos, thiophanate methyl, carboxin etc are also found to be effective.

Rusts

14. Tropical rust

Symptoms

» Outbreaks of this rust are sporadic and confined to the American tropics.

» Pustules are varying in shape from round to oval.

» They are small and found beneath the epidermis. At the center of the pustule the lesion appears white to pale yellow and an opening develops.

» The pustule is sometimes black rimmed, but its center remains light.

Fungus: *Phakopsora zeae* (Mains) Buriticá
Syn: *Angiopsora zeae* Mains; *Physopella zeae* (Mains) Cummins & Ramachar

Diseases of Maize

Fungal characters

Uredospores are globose or elliptical finely echinulate, yellowish brown with 4 germpores. Teliospores are brownish black, or dark brown, oblong to ellipsoidal, rounded to flattened at the apex. They are two celled and slightly constricted at the septum and the spore wall is thickened at the apex.

Disease cycle

Primary source of inoculums is uredospores surviving on alternate hosts *viz., Oxalis corniculata* and *Euchlaena mexicana.*

Favourable conditions

» Cool temperature and high relative humidity.

Management

» Grow resistant varieties like CM 105, CM 111, OH 43, OH 545, CSH 1, CSH 2 and N 28.

» Remove the alternate hosts.

» Spray mancozeb at 2 kg/ha.

15. Common rust: *Puccinia sorghi* Schwein.

» The disease is found worldwide in subtropical, temperate, and highland environments with high humidity.

» Circular to oval, elongated cinnamon-brown powdery pustules are scattered over both surface of the leaves.

» As the plant matures, the pustules become brown to black owing to the replacement of red uredospores by black teliospores.

» Plants of the alternate host (*Oxalis* spp.) are frequently infected with light orange colored pustules.

16. Southern rust: *Puccinia polysora* Underw.

» Pustules are smaller, lighter in color (light orange), and more circular than those produced by *P. sorghi.*

90   Diseases of Field Crops and their Management

» They are also present on both leaf surfaces, but the epidermis remains intact longer than it does in *P. sorghi* infected leaves.

» Pustules turn dark brown as plants approach maturity.

» *Polysora* rust (or southern rust) is common in hot and humid lowland tropical conditions.

Smut

17. Common smut

Symtoms

» Galls develops on ears, leaves, stalk, or tassels.

» Galls initially are covered with white to silvery tissue.

» Interior of galls develop into a dark mass of dark spores.

» This pathogen is present where corn is grown around the world.

Fungus: *Ustilago maydis* (DC.) Corda
      Syn: *Ustilago zeae* (Beckm.) Unger).

18. Head smut

Symptoms

» Symptoms are usually noticed on the cob and tassel.

» Large smut sori replace the tassel and the ear.

» Sometimes the tassel is partially or wholly converted into g sorus.

» The smutted plants are stunted produce little yield and remain greener than that of the rest of the plants.

» Smut spores are produced in large numbers which are reddish brown to black, thick walled, finely spined, spherical.

Fungus: *Sporisorium reilianum* J.G. Kühn Langdon & Full.
      Syn: *Sorosporium reilianum* (J. G. Kühn) McAlpine;
      Sphacelotheca reiliana (J. G. Kühn) G. P. Clinton;

Diseases of Maize

Sporisorium holci-sorghi (Rivolta) Vánky;
Sporisorium reilianum (J. G. Kühn) Langdon & Full. f. sp. *zeae* (J.G. Kühn);
Ustilago reiliana J. G. Kühn

19. False head smut

Symptoms

» The symptoms of the disease as observed were transformation of individual male flowers of the tassel into dark green spore (sori) balls with velvety appearance.

» The interiors of the false smut balls were interwined with hyphae at early stage and then chlamydospores are formed out of these mycelia.

» The chlamydospores are almost smooth when young and become warty when mature.

Fungus: *Ustilaginoidea virens* (Cooke) Takah
Syn: *Ustilaginoidea oryzae* (Pat.) Bref.

The warty and olivaceous spores are borne on minute sterigmata that project from radial hyphae.

Disease cycle

The smut spores retain its viability for two years. The fungus is externally seedborne and soil-borne. The major source of infection is through soil-borne chlamydospores.

Favourable conditions

» Low temperature favours more infection and this fungus also infects the sorghum

Management

» Field sanitation.

» Crop rotation with pulses.

» Seed treatment with captan at 4 g/ kg.

20. Charcoal post flowering stalk rot

Symptoms

» The affected plants exhibit wilting symptoms.

» The stalk of the infected plants can be recognized by grayish streak.

» The pith becomes shredded and grayish black minute sclerotia develop on the vascular bundles.

» Shredding of the interior of the stalk often causes stalks to break in the region of the crown.

» The crown region of the infected plant becomes dark in colour.

» Shredding of root bark and disintegration of root system are the common features.

Fungi: *Macrophomina phaseolina* (Tassi) Goid.
Syn: *Botryodiplodia phaseoli* (Maubl.) Thirum.;
Macrophomina phaseoli (Maubl.) S. F. Ashby; *Sclerotium bataticola* Taubenh.

The fungus produces large number of sclerotia which are round and black in colour. Sometimes, it produces pycnidia on the stems or stalks.

Disease cycle

The fungus has a wide host range, attacking sorghum, pearlmillet, fingermillet and pulses. It survives for more than 16 years in the infected plant debris. The primary source of infection is through soil-borne sclerotia. The pathogen also attacks many other hosts, which helps in its perpetuation. Since the fungus is a facultative parasite it is capable of living saprophytically on dead organic tissues, particularly many of its natural hosts producing sclerotial bodies. The fungus over winters as a sclerotia in the soil and infects the host at susceptible crop stage through roots and proceeds towards stem.

Favourable conditions

» High temperature and low soil moisture (drought).

Management

» Long crop rotation with crops that are not natural host of the fungus.

Diseases of Maize

» Residue management through sanitation, tillage and ploughing down of crop debris

» Grow disease tolerant varieties *viz.*, JH 6805, Bio 9636, Pusa early hybrid, X 1280, JKMH-1701, JH 6805, Bio 9639, Bio 9636, X 1280, CoH (M) 5, Ganga 5, Ganga 101, Amber, Jawahar, Ganga Safed 2, DHM105, SN-65, SWS-8029, Diva and Zenit.

» Seed treatment with carbendazim or captan at 2 g/kg.

» Balanced soil fertility; avoid high level of N and low level of K.

» Manage the attack of borers in maize as their injury predisposes to stalk rot.

» In stalk rot affected field, balance soil fertility specially increases the potash level up to 80 kg/ ha help in minimising the disease.

» Avoiding water stress at flowering time, earhead emergence and maturity.

» Apply *Trichoderma asperellum* in furrows after mixing with FYM at 10 g/ kg FYM (1 kg/100 kg FYM /acre) at least 10 days before its use in the field in moist condition.

21. *Pythium* stalk rot

Symptoms

» Usually the basal internodes become soft, dark brown water soaked, causing the plants lodge.

» Damaged internodes commonly twist before the plants lodge.

» Diseased plants can remain alive until all vascular bundles become affected.

» Isolations in culture media are necessary to differentiate *Pythium* from *Erwinia* stalk rots.

Chromista: *Pythium aphanidermatum* (Edson) Fitzp.
Syn: *Pythium butleri* Subram.

Management

» Maintain plant population around 50,000/ ha.

» Good field drainage.

» Removal of previous crop debris.

94   Diseases of Field Crops and their Management

- » Soil drench with captan at basal internode (5-7week growth stage) at 1g/ lit of water.

- » Grow resistant varieties like Ganga and Safed 2.

22. *Fusarium* stalk rot and wilt

Symptoms

- » Affected plant wilts, leaves change from light to dull green, and the lower stalks become straw coloured.

- » Reddish discoloration occurs inside the infected stalk.

- » The internal pith tissue disintegrates, leaving only the vascular bundles.

- » Fungus enters through roots and grows up in to lower stem.

- » If infection occurs just after flowering, husks appear bleached and straw coloured.

Fungi: *Fusarium* spp.

- » *Fusarium proliferatum* (Matsush.) Nirenberg ex Gerlach & Nirenberg
  Syn: *Gibberella fujikuroi* (Sawada) Ito, mating population D;
  Gibberella intermedia (Kuhlman) Samuels, Nirenberg & Seifert)

- » *Fusarium subglutinans* (Wollenw. & Reinking) P.E. Nelson,
  Syn: *Gibberella fujikuroi* (Sawada) Ito, mating population E;
  Gibberella subglutinans (E.T.Edwards) P. E. Nelson, Toussoun & Marasas

- » *Fusarium temperatum* Scaufl. & Munaut

- » *Fusarium verticillioides* (Sacc.) Nirenberg
  Syn: *Fusarium moniliforme* J. Sheld.; *Gibberella moniliformis* Wineland
  Gibberella fujikuroi (Sawada) Ito, mating population A;

Management

- » Application of potassic fertilizers reduces infection.

- » Seed from infected areas should not be planted.

- » Rotation with other crops.

Diseases of Maize

» Solarization, fumigation and soil drenches with bioagents and fungicides to mitigate the soil inoculum.

» Grow resistant varieties like Pusa early hybrid, X 1280, Ranjit and Ganga 5

» Enrich the soil with *Trichoderma asperellum* impregnated composts for good development of antagonists against the soil borne pathogens.

» Application of captan 75% at 12 g/ 100 litre of water as soil drench at the base of the plants when crop is 5 to 7 week old.

23. Brown spot

It was first noticed in Bihar by Shaw in 1910. Since it occurs sporadically in mild form, causing little or no damage, it is of minor importance. The fungus infects teosinte also. The disease is common in low lying and ill-drained fields.

Symptoms

» The first noticeable symptoms develop on leaf blades and consist of small chlorotic spots, arranged as alternate bands of diseased and healthy tissue.

» Spots on the mid-ribs are circular and dark brown, while lesions on the laminae continue as chlorotic spots.

» Nodes and internodes also show brown lesions. In severe infections, these may coalesce and induce stalk rotting and lodging.

Fungus: *Physoderma maydis* (Miyabe) Miyabe
   Syn: *Physoderma zeae-maydis* F. J. F. Shaw

Management

» Planting corn early allow to escape infection.

» Residue management through sanitation, tillage and ploughing down of crop debris.

» Follow crop rotation with non-host crops.

» Removing of *Saccharum spontaneoum* grass growing around the crop, can minimise the disease.

» Grow resistant varieties and hybrids like Ganga 11, Deccan, Deccan 103, Composite

96   Diseases of Field Crops and their Management

Suwan1, F-9572 A, JKMH 178-4, FH 3113, JH 10655, FH 3113, DMR 1 and DMR 5.

» Spraying of copper oxychloride 0.25% or metalaxyl 0.1%.

24. *Curvularia* leaf spot

Symptoms

» The disease appears in the form of flecking which later develops into larger lesions.

» The lesions in general are round to oval, separate or coalescent, 1-6 mm in diameter.

» The centre of each lesion is straw coloured to light brown, which is surrounded by a dark brown margin.

Fungus: *Curvularia aeria* (Bat., J.A. Lima & C.T. Vasconc.) Tsuda
         Syn: *Curvularia lunata* (Wakker) Boedijn var. aeria

Management

» Seed treatment with thiram or captan at 2 g/ kg reduces seed infection.

» Two sprayings of captafol at 2 g/ lit of water.

25. Grey leaf spot

Fungi: *Cercospora zeae-maydis* Tehon & E. Y. Daniels
        Cercospora zeina Crous & U. Braun

Symptoms

» Initial lesions appear as greenish black water soaked circular areas with chlorotic halos.

» Later expanding into oval and then the diagnostic parallel side rectangular brownish gray lesions.

Management

» Crop rotation with leguminous crop.

» Grow resistant variety like EVS 4.

Diseases of Maize

> Seed treatment with thiram or captan at 2 g/ kg reduces seed infection.

> Two sprayings of captafol at 2 g/ lit of water.

26. Anthracnose

Fungus: *Colletotrichum graminicola* (Ces.) G.W. Wills.

Symptoms

> Irregular, oval to elongated spot with tapered ends.

> The spots are water soacked first later but later on become brown with red brown margin.

> The lesions may be 15 mm in dia and coalesce before withering and result in death of the affected leaves.

27. *Rostratum* leaf spot

Symptoms

> Infected leaves show small, elongated, pale, yellow spots.

> These spots elongated gradually to form longitudinal stripes between the veins.

> In severe cases the lesions may coalesce and extend across the veins having 2-3 mm.

Fungus: *Exserohilum rostratum* (Drechsler) K. J. Leonard & Suggs
  Syn: *Bipolaris rostrata* (Drechsler) Shoemaker;
  Drechslera rostrata (Drechsler) M.J.Richardson & E.M.Fraser;
  Exserohilum halodes (Drechsler) K.J. Leonard & Suggs;
  Helminthosporium rostratum Drechsler;
  Setosphaeria rostrata K.J. Leonard

28. Bacterial stalk rot

Symptoms

> The basal internodes develop soft rot and give a water soaked appearance.

> A mild sweet fermenting odour accompanies such rotting.

98 Diseases of Field Crops and their Management

» Leaves some time show signs of wilting and affected plants topple down in few days.

» Ears and shank may also show rot.

» They fail to develop further and the ears hang down simply from the plant

Bacteria: *Erwinia dissolvens*; *Erwinia chrysanthemi* pv. *zeae*; *Dickeya zeae*; *Erwinia carotovora* f. sp. *zeae* Sabet; *Pectobacterium chrysanthemi* pv. *zeae* Kelman

Disease cycle

Borer insects play a significant role in initiation of the disease. The organism is soil borne and makes its entry through wounds and injuries on the host surface. The organism survives saprophytically on debris of infected materials and serves primary inoculum in the next season.

Management

» Grow resistant varieties like PAU 352, PEMH 5, DKI 9202 and DKI 9304.

» Grow tolerant cultivar Ganga 2.

» Soil drench with bleaching powder (chlorine 33%) at 10 kg/ ha as at pre-flowering stage.

» Planting crop on ridges.

» Avoid water logging and proper drainage.

» Planting of the crop on ridges rather than flat soil.

» Avoid use of sewage water for irrigation.

» Sprinkler irrigation of chlorinated water help in control of bacterial stalk rot of maize.

» Spray klorosin 25 kg/ ha as soil drenching at flowering stage.

29. *Maize dwarf mosaic virus*

Symptoms

» Often begin as chlorotic spots and streaks on green, young leaves, which later develop into a mottle or a mosaic pattern.

Diseases of Maize 99

- » Viral strain, corn genotype, and stage of corn development at the time of infection will affect the type of symptoms.

- » Upper internodes of corn may be shortened, and excessive tillering may occur.

- » Ear formation and development may slow, which may cause grain yield loss.

- » Hybrids infected early in their growth stage may be stunted.

Transmission

The corn leaf aphid (*Rapalosiphum maidis*) and the green peach aphid (*Myzus persicae*) are the most common vectors of MDMV. Also, can be seed transmitted at a low frequency or mechanically transmitted by leaf rubbing, etc. survives in Johnson grass and sorghum.

30. *Maize mosaic virus I (Potyvirus)*

Symptoms

- » Symptoms appear as chlorotic spots, which gradually turn into stripes covering entire leaf blade.

- » Chlorotic stripes and spots can also develop on leaf sheaths, stalks and husks.

- » Moderate to severe rosetting of new growth is observed.

- » Size of stalk, leaf blades and tassel tend to be normal in late infection.

Virus: It is caused by *Maize mosaic virus I*. Virions are flexuous, 750-900nm long, ssRNA genome.

Transmission: It is transmitted in nature by leaf hopper vector, *Perigrimus maidis.*

31. *Sugarcane mosaic virus*

- » *Sugarcane mosaic virus* is also known to infect maize, causing characteristic mosaic symptoms.

- » It is common on sugarcane, sorghum and maize, but the damage to maize is much less than that caused by the maize mosaic virus.

32. *Maize streak virus (Geminivirus)*

Symptoms

- » Disease symptoms first become apparent about a week following infection.

- » Small (0.1 to 2 mm in diameter), chlorotic, circular spots arise on the basal sections of young leaves that emerge after infection.

- » Spots contrast sharply with surrounding healthy green leaf tissue.

- » Numbers of spots increase with plant growth and can coalesce.

- » In susceptible varieties, spots enlarge parallel to leaf veins, forming distinctive elongated, chlorotic streaks distributed evenly over the leaf surface. Infection with MSV commonly results in crop stunting and barren plants.

Transmission: The leafhopper *Cicadulina mbila* is the most important insect vector of MSV.

Minor diseases

- » **Leaf spot**: *Tramatosphaeria maydis* (P. Henn.) Rang. *et al.*

- » **Leaf spot**: *Didymella exitialis* (Mor.) Millier.

- » **Leaf spot**: *Diplodia macrospora* Earle.

- » **Leaf spot**: *Hypochnus sasalcii* Shirai.

- » **Zonate spot**: *Gloeocercospora sorghi* Bain Edg.

- » **Ear rot**: *Cephalosporium acremonium* Corda; *Rhizoctonia zeae* Voorhees.

- » **Maize wilt:** *Pantoea stewartii* subsp. *stewartii*

- » **Bacterial stalk rot**: *Pseudomonas syringae* pv. *lapsa*

- » **Bacterial stalk rot**: *Erwinia carotovora* sub. sp. *dissolvens*

- » **Bacterial stalk and ear rot**: *E. carotovora* sub. sp. *zeae*

- » **Bacterial leaf spot**: *Xanthomonas campestris* pv. *maydis*

- » **Bacterial leaf stripe**: *X. campestris* pv. *rubrilineans* (Lee *et al.*) Starr and Burk.

- » *Maize chlorotic dwarf virus* **(MCDV)**

- » *Maize chlorotic mottle virus* **(MCMV)**

Diseases of Maize

» ***Maize lethal necrosis* (MLN)**

» ***Maize stripe virus* (M StV)**

» ***Maize rough dwarf virus* (MRDV)**

» ***Maize fine stripe virus***

» ***Maize bushy stunt* (MBS)**

» **Corn stunt:** *Spiroplasma kunkelii*

Chapter - 6

Diseases of Sorghum/ Jower - *Sorghum bicolor* (L.) Moench

S. No.	Disease	Pathogen
1.	Sorghum downy mildew	*Peronosclerospora sorghi*
2.	Crazy top	*Sclerophthora macrospora*
3.	Leaf blight	*Setosphaeria turcica*
4.	Rectangular grey leaf spot	*Cercospora sorghi*
5.	Anthracnose and red rot	*Colletotrichum graminicola*
6.	Rust	*Puccinia purpurea*
7.	Grain smut	*Sphacelotheca sorghi*
8.	Loose smut	*Sporisorium cruentum*
9.	Long smut	*Tolyposporium ehrenbergii*
10.	Head smut	*Sporisorium reilianum*
11.	Ergot	*Sphacelia sorghi*
12.	Head mould	Fungal complex
13.	Charcoal rot	*Macrophomina phaseolina*
14.	Bacterial leaf stripe	*Pseudomonas andropogonis*
15.	Bacterial leaf streak	*Xanthomonas vasicola* pv. *holcicola*
16.	Witch weed	*Striga asiatica*

Diseases of Sorghum 103

17.	Maize stripe disease	*Maize stripe virus*
18.	Maize mosaic disease	*Maize mosaic virus*
19.	Red stripe disease	Virus

Downy mildew

1. Sorghum downy mildew / Leaf shreading

Symptoms

» The pathogen causes localized and systemic downy mildew of sorghum.

» It invades the growing points of young plants, either through oospore or conidial infection. As the leaves unfold they exhibit green or yellow colouration.

» First symptoms are visible on the lower part of the leaf blade, which later progress upward. Abundant downy white growth is produced on the lower surface of the leaves, which consists of sporangiophores and sporangia.

» Normally three or four leaves develop the chlorotic downy growth. Subsequent, leaves show progressively more of a complete bleaching of the leaf tissue in streaks or stripes.

» As the infected bleached leaves mature they become necrotic and the interveinal tissues disintegrate, releasing the resting spores (oospores) and leaving the vascular bundles loosely connected to give the typical shredded leaf symptom.

» Finally, the infected plants shows stunting and sterility.

Chromista: *Peronosclerospora sorghi* (W.Weston & Uppal) C.G. Shaw;
Syn: *Sclerospora sorghi* W. Weston & Uppal

2. Crazy top

» The symptoms vary considerably with the host cultivar.

» Young sorghum plants may develop mottled leaves resembling mosaic virus infection and subsequently develop thick, stiff, twisted or curled leaves.

» Yellowing occurs later.

» Oospores are formed in the infected leaf tissues, but the leaf shredding characteristic does not occur with this disease.

» Infected plants tiller profusely, may not head or the heads may be barren or they may show proliferation of the floral structures.

Chromista: *Sclerophthora macrospora* (Sacc.) Thirumalachar *et al*.
Syn: *Sclerospora macrospora* Sacc.

It is an obligate parasite, systemic in young plant. The mycelium is intercellular, non-septate. Sporangiophores emerge through the stomata in single or in clusters which are stout and dichotomously branched. Spores are single celled, hyaline, globose and thin walled. Oospores are spherical, thick walled and deep brown in colour.

Disease cycle

The primary infection is by means of oospores present in the soil which germinate and initiate the systemic infection. Oospores persist in the soil for several years. Secondary spread is by air-borne sporangia. Presence of mycelium of the fungus in the seeds of systemically infected plants is also a source of infection. The disease has been known to occur through a collateral host, *Heteropogen centortus* on which the fungus perpetuates of the host. The breakdown of tissue causes shredding. The oospores either fall to the soil or are wind blown, often within host tissue. They can remain viable in the soil for 5-10 years. Conidia are formed at night in large numbers. The optimum temperature for production is 20-23°C.

Favourable conditions

» Maximum sporulation takes place at 100% relative humidity.

» Optimum temperature for sporulation is 21-23°C during night.

» Light drizzling accompanied by cool weather is highly favourable.

Management

» Crop rotation with other crops *viz.*, pulses and oilseeds.

» Avoid the secondary spread of the disease by roguing out the infected plants since the wind plays a major role in the secondary spread of the disease.

» Soil spreading with potassium azide at 1.12 kg/ ha.

Diseases of Sorghum 105

- » Grow moderately resistant varieties like Co 25, Co 26, CSH 2, CSV 5, SPV 101 and hybrid SPH 1705.

- » Seed treatment with metalaxyl at 6 g/ kg.

- » Spray chytrid fungus (*Gaertneriomyces* sp.) parasitizing oospores of *P. sorghi*.

- » Spray metalaxyl 500 g or mancozeb 2 kg or ziram 1 kg or zineb 1 kg/ ha.

- » Alternatives, like phosphonic acid may be useful in managing the disease where metalaxyl resistance is a problem.

3. Leaf blight

Symptoms

- » The pathogen also causes seed rot and seedling blight of sorghum.

- » The disease appears as small narrow elongated spots in the initial stage and in due course they extend along the length of the leaf.

- » On older plants, the typical symptoms are long elliptical necrotic lesions, straw coloured in the centre with dark margins.

- » The straw coloured centre becomes darker during sporulation.

- » The lesions can be several centimeters long and wide.

- » Many lesions may develop and coalesce on the leaves, destroying large areas of leaf tissue, giving the crop a burnt appearance.

Fungus: *Setosphaeria turcica* (Luttrell) K. J. Leonard & E. G. Suggs
Syn: *Exerohilum turcicum* (Pass.) K. J. Leonard & E. G. Suggs
Syn: *Helminthosporium turcicum* (Pass)

The mycelium is localised in the infected lesion. Conidiophores emerge through stomata and are simple, olivaceous, septate and geniculate. Conidia are olivaceous brown, 3-8 septate and thick walled.

Disease cycle

The pathogen is found to persist in the infected plant debris. Seed borne conidia are responsible for seedling infection. Secondary spread is through wind-borne conidia.

Favourable conditions

» Cool moist weather, high humidity 90% and high rainfall.

Management

» Use disease free seeds.

» Grow resistant varieties like Tift Sudan, Co 21, Co 22, Co 23, CSH 6 and CSV 6.

» Seed treatment with captan at 4 g/ kg.

» Spray mancozeb 1.25 kg or captafol 1 kg/ ha.

4. Rectangular grey leaf spot

Symptoms

» The symptoms appear as small leaf spots which enlarge to become rectangular lesions (which can be 5-15 mm long by 2 to 5 mm wide) on the leaf and leaf sheath.

» Usually the lower leaves are first attacked.

» The lesions are typical dark red to purplish with lighter centers.

» The lesions are mostly isolated and limited by veins.

» The colour of the spots varies from red, purple, brown or dark depending upon the variety.

Fungus: *Cercospora sorghi* Ellis & Everh.

Mycelium of the fungus is hyaline and septate. Conidiophores emerge in clusters through stomata, which are brown and simple, rarely branched. Conidia are hyaline, thin walled, 2-13 celled and long obclavate.

Disease cycle

The conidia survive up to 5 months. The disease spreads through air-borne and seed-borne conidia.

Favourable conditions

» Cool moist weather, high humidity (90%) and high rainfall.

Diseases of Sorghum

Management

» Use disease free seeds.

» Grow resistant varieties like Co 21, Co 22, Co 23, CSH-6 and CSV-6.

» Seed treatment with captan at 4 g/kg.

» Spray mancozeb 2 kg/ ha or chlorothalanil 2 kg/ ha or trifloxystrobin + tebuconazole 2 kg/ ha.

5. Anthracnose and red rot

Symptoms

» The fungus causes both leaf spot (anthracnose) and stalk rot (red rot).

» The disease appears as small red coloured spots on both surfaces of the leaf.

» The centre of the spot is white in colour encircled by red, purple or brown margin.

» Numerous small black dots like acervuli are seen on the white surface of the lesions.

» Red rot can be characterized externally by the development of circular cankers, particularly in the inflorescence.

» Infected stem when split open shows discoloration, which may be continuous over a large area or more generally discontinuous giving the stem a marbeled appearance.

Fungus: *Colletotrichum graminicola* (Ces.) G. W. Wilson
Syn: *Glomerella graminicola* Politis

The mycelium of the fungus is localised in the spot. Acervuli with setae arise through epidermis. Conidia are hyaline, single celled, vacuolate and falcate in shape.

Disease cycle

The disease spread by means of seed-borne and air-borne conidia and also through the infected plant debris and collateral hosts Sudan grass and Johnson grass.

Favourable conditions

» Continuous rainfall, temperature of 28-30°C and high humidity.

Management

>   Seed treatment with captan at 4 g/ kg.

>   Grow resistant varieties Texas Milo, Tift Sudan, CSV 17 and SPV 162.

>   Grow resistant lines IS 2095, ICSB 12021, IS 10302, IS 23521 and IS 473.

>   Spray the crop with mancozeb 2 kg/ ha or chlorothalanil 2 kg/ ha.

6.   Rust

Symptoms

>   The fungus affects the crop at all stages of growth.

>   The first symptoms are small flecks on the lower leaves (purple, tan or red depending upon the cultivar).

>   Pustules (uredosori) appear on both surfaces of leaf as purplish spots which rupture to release reddish powdery masses of uredospores.

>   Teliopores develop later sometimes in the old uredosori or in telisori, which are darker and longer than the uredosori.

>   The pustules may also occur on the leaf sheaths and on the stalks of inflorescence.

Fungus: *Puccinia purpurea* Cooke

The uredospores are pedicellate, elliptical or oval, thin walled echinulated and darkbrown in colour. The teliospores are reddish or brown in colour and two celled, rounded at the apex with one germ pore in each cell. The teliospores germinate and produce promycelium and basidiospores. Basidiospores infect *Oxalis corniculata* (alternate host) where pycnial and aecial stages arise.

Disease cycle

The air-borne uredospores survive for a short time in soil and infected debris. Presence of alternate host helps in perpetuation of the fungus.

Favourable conditions

>   Low temperature of 10 to 12°C favours teliospore germination.

Diseases of Sorghum

» A spell of rainy weather favours the onset of the disease.

Management

» Grow resistant varieties like CSV 5, CSV 17, IS 3443, IS 3547, IS 18758 and SPV-247.

» Remove the alternate host *Oxalis corniculata.*

» Spray the crop with mancozeb at 2 kg/ ha or sulphur 25 kg/ ha.

7. Grain smut

Other names: Kernel smut, Covered smut, Short smut

Symptoms

» The individual grains are replaced by smut sori.

» The sori are oval or cyclindrical and are covered with a tough creamy skin (peridium) which often persists unbroken up to thrashing.

» Ratoon crops exhibited higher incidence of disease.

Fungi: *Sphacelotheca sorghi* (Link) G. P. Clinton; *Sporisorium sorghi* Link in Willd

Disease cycle

» The pathogen is externally seed borne.

Management

» Grow resistant varieties like CSH 5, CSH 9, SDM 9, SPV 102, SPV 115, SPV 138 and SPV 245.

» Seed treatment with Agrosan 2 g/ kg or fine powdered sulphur 5 g/ kg.

» Avoid ratooning.

8. Loose smut/ Kernel smut

Symptoms

» The affected plants can be detected before the ears come out.

» They are shorter than the healthy plants with thinner stalks and marked tillering.

» The ears come out much earlier than the healthy.

» The glumes are hypertrophied and the earhead gives a loose appearance than healthy.

» The sorus is covered by a thin membrane which ruptures very early, exposing the spores even as the head emerges from the sheath.

Fungus: *Sporisorium cruentum* (Kuhn) K. Vanky
Syn: *Sphacelotheca cruenta* (Kuhn) A. A. Potter

Disease cycle

» The pathogen is externally seed borne.

Favourable conditions

» Temperatures between 20-25°C and slightly acidic soils.

Management

» Collect smutted ear-heads in cloth bags and destroy by dipping in boiling water or bury in soil.

» Avoid ratooning.

9.   Long smut

Symptoms

» This disease is normally restricted to a relatively a small proportion of the florets which are scattered on a head.

» The sori are long, more or less cylindrical, elongated, slightly curved with a relatively thick creamy-brown covering membrane (peridium).

» The peridium splits at the apex to release black mass of spores (spore in groups of balls) among which are found several dark brown filaments which represent the vascular bundles of the infected ovary.

Fungi: *Tolyposporium ehrenbergii* (J.G. Kuhn) Pat
Syn: *Anthracocystis ehrenbergii* (J. G. Kuhn) McTaggart & R.G. Shivas
Syn: *Sporisorium ehrenbergii* K. Vánky

Diseases of Sorghum

Disease cycle

- » The spore balls are soil borne.

- » The sporidia produced by soil borne spore ball become wind borne to the buds and intiate a systemic infection.

Management

- » Seed treatment with captan or thiram at 4 g/ kg.

- » Use disease free clean seeds and follow crop rotation.

- » Collect the smutted ear heads in cloth bags and bury in soil.

10. Head smut

Symptoms

- » The entire head is replaced by large sori.

- » The sorus is covered by a whitish grey membrane of fungal tissue, which ruptures, before the head emerges from the boot leaf to expose a mass of brown smut spores.

- » Spores are embedded in long, thin, dark colored filaments which are the vascular bundles of the infected head.

Fungus: *Sporisorium reilianum* (Kühn) Langdon & Fullerton
Syn: *Sphacelotheca reiliana* (Kühn) G. P. Clinton;
Syn: *Sporisorium holci-sorghi* (Rivolta) K. Vánky

Disease cycle

- » The smut spores retain viable for several years. The pathogen is externally seed borne and soil borne.

Favourable conditions

- » Infection on seedlings takes place at 21-28°C.

Management

- » Treat the seed with captan at 4 g/ kg.

112   Diseases of Field Crops and their Management

»   Use disease free seeds and follow crop rotation.

»   Collect the smutted ear heads in cloth bags and bury in soil.

Comparison of the Characters of four Smuts of Sorghum found in India

Characters	Grain smut	Loose smut	Long smut	Head smut
Organisms	*S. sorghi*	*S. cruenta*	*T. ehrenbergii*	*S. reiliana*
Ear infection	All or most grains smutted	All or most grains smutted	Only about 2 % grains are infected	Entire head smutted into a single sorus
Sori	Small, 5-15 × 3-5 mm	Small, 3-18 × 2-4 mm	Long, 40 mm × 6-9 mm	7.5-10 cm × 2.5-5 cm
Columella	Short columella present	Long columella present	Columella absent, but 8-10 vascular strands present	Columella absent but a network of vascular tissue present
Spores	Single, round to oval, olive brown, smooth walled, 5-9 μ in diameter	Single, sperical or elliptical, dark brown, spore walls pitted, 5-10 μ in diameter	Always in balls, globose or angular, brownish green, warty spore wall, 12-16 μ in diameter	Loosely bound into balls, spherical or angular, dull brown, minutely papillate 10-16 μ in diameter
Viability of spores	Over 10 years	About 4 years	About 2 years	Up to 2 years
In culture	Yeast-like growth with sporidia	In colonies with sporidia and resting spores 40-50 μ in diameter	In colonies with mass of sporidia	In colonies germtubes and sporidia
Spread	Externally seed borne	Externally seed borne	Air-borne	Soil-borne and seed borne

11.   Ergot / Sugary disease

Symptoms

»   The disease is confined to individual spikelets.

»   The first symptom is the secretion of honey dew from infected florets.

»   Under favourable conditions, long, straight or curved, cream to light brown, hard sclerotia develop.

»   Often the honey dew is colonised by *Cerebellum sorghi-vulgaris* (a parasitic fungi of *Sphacelia*) which gives the head a blackened appearance.

Diseases of Sorghum

Fungi: *Sphacelia sorghi* McRae (Asci stage); *Claviceps sorghi* Kulkarni (Sclerotial stage)

The fungus produces septate mycelium. The honey dew is a concentrated suspension of conidia, which are single celled, hyaline, elliptic or oblong.

Disease cycle

The primary source of infection is through the germination of sclerotia which release ascospores that infect the ovary. The secondary spread takes place through air and insect-borne conidia. Rain splashes also help in spreading the disease.

Favourable conditions

- » A period of high rainfall and high humidity during flowering season.
- » Cool night temperature and cloudy weather aggravate the disease.

Management

- » Adjust the date of sowing so that the crop does not flower during September- October when high rainfall and high humidity favor the disease.
- » Grow resistant varieties like Nandiyal, Bilichigan, SPV 191 and CSH 5.
- » Follow crop rotation with non-host crops.
- » Spray mancozeb 2 kg/ ha or carbendazim at 500 g/ ha at emergence of ear head (5-10 % flowering stage) followed by a spray at 50% flowering and repeat, if necessary.

12.   Head mould/ Grain mould/ Head blight

More than thirty two genera of fungi were found to occur on the grains of sorghum.

Symptoms

- » If rain occurs, during flowering and grain filling stages, severe grain mould was takes place.
- » Symptom varies depending upon the organism involved and the degree of infection.
- » The most frequently occurring genera are *Fusarium, Curvularia, Alternaria, Aspergillus* and *Phoma. Fusarium semitectum* and *F. moniliforme* develop a fluffy white or pinkish coloration. *C. lunata* colours the grain black.

Disease cycle

The fungi mainly spread through air-borne conidia. The fungi survive as parasites as well as saprophytes in the infected plant debris.

Favourable conditions

- » Wet weather following the flowering favors grain mould development.
- » The longer wet period the greater the mould development.
- » Compact ear heads are highly susceptible.

Management

- » Adjust the sowing time.
- » Grow resistant varieties like GMRP 4, GMRP 9, GMRP 13, DSR-GMN 41, DSR-GMN 42, DSR-GMN 46, DSR-GMN 52, DSR-GMN 58, DSR-GMN 59, SPV 2205 and tolerant varieties like CSV 15
- » Spray mancozeb 1 kg/ ha or captan 1 kg + Aureofungin-sol 100 g/ ha. in case of intermittent rainfall during earhead emergence, a week later and during milky stage.

13. Charcoal rot / Hollow stem / Stalk rot / Blight

Symptoms

- » Sudden wilting and death of the diseased plant resulting in lodging.
- » If the infected stalk is split open, the pith is disintegrated with longitudinal shredding of the tissue into fibers.
- » Small black sclerotial bodies are seen in the infected tissues.
- » The stem, breaks near the ground level. Premature ripening takes place and the heads are poorly developed.

Fungus: *Macrophomina phaseolina* (Tassi) Goidanich

Disease cycle

- » Primary infection: Sclerotial bodies present in the soil, infected debris and weed host

Diseases of Sorghum

» Secondary infection: Sclerotial bodies carried through rain or irrigated water.

Favourable conditions

» Soil temperature 35°C

» Moisture stress conditions preceding crop maturity

» High dose of nitrogenous fertilizers

Management

» Grow resistant varieties like E-36 1, CSV 5, CSH 7-R, SPV 126 and SPV 193.

» Minimize the plant population (60,000 plants/ ha).

» Collected and burning of infected plants along with trash.

» Avoid moisture stress at flowering.

14. Bacterial leaf stripe

Bacterium: *Pseudomonas andropogonis* Smith (Stapp).
Burkholderia andropogonis (Smith) Gillis *et al.*

Symptoms

» Initial symptoms are small (1 cm long), linear, intervenal lesions.

» Lesions on leaves and sheaths are purple, red, yellow or tan, depending on the host reaction.

» Water soaking of tissue adjacent to a lesion is usually not observed under field conditions.

» Bacterial exudates are usually observed from infected portions of the leaf under microscopic observation.

» Lesions may also occur on the kernel, peduncle, and rachis, and also in the pith of the stalk.

15. Bacterial leaf streak or bacterial streak

Bacterium: *Xanthomonas vasicola* pv. *holcicola* Elliott.

Symptoms

> » First symptoms are narrow, water-soaked, transparent leaf streaks, 2-3 mm wide by 2-15 mm long, appearing as early as the second leaf stage of the seedling.

> » Lesions soon turn red, become opaque, and at intervals may broaden into somewhat irregularly shaped oval spots with tan centers and narrow red margins.

> » In severe attacks, these coalesce to form long irregular streaks and blotches extending across all or much of the leaf blade, with dead tissue bordered by narrow, dark margins between the reddish-brown streaks.

> » Abundant bacterial exudates are produced as light-yellow droplets, which dry to thin white or cream scales.

16. Phanerogamic parasite: *Striga asiatica* (L.) Kuntze (Asian orgin)

S. densiflora (African orgin)
S. forbesi Benth.
S. gesnerioides (Willd.) Vatke
S. hermonthica (Del.) Benth.

It is a partial root parasite and occurs mainly in the rainfed sorghum. It is a small plant with bright green leaves, grows up to a height of 15-30 cm. The plants occur in clusters of 10-20/ host plant. *S. asiatica* produces red to pink flowers while. *S. densiflora* produces white flowers. Each fruit contains minute seeds in abundance which survives in the soil for several years.

The root exudates of sorghum stimulate the seeds of the parasite to germinate. The parasite then slowly attaches to the root of the host by haustoria and grows below the soil surface producing underground stems and roots for about 1-2 months. The parasite grows faster and appears at the base of the plant. Severe infestation causes yellowing and wilting of the host leaves. The infected plants are stunted in growth and may die prior to seed setting.

Management

> » Regular weeding and intercultural operation during early stages of parasite growth.

> » Hand weeding of the parasites before flowering.

> » Crop rotation with cowpea, groundnut and sunflower.

Diseases of Sorghum

» Grow resistant varieties like Co 20.

» Mixing of Ethrel with soil triggers germination of *Striga* in the absence of host.

» Spray fernoxone (sodium salt of 2,4-D) or agroxone (MCPA) at 450g /500 liters of water or paraquat at 1 kg/ ha.

» Spray tetrachloro dimethyl phenoxy acetate 1% can be used for instant killing of *Striga*, if water is in scarce.

Virus diseases of Sorghum

Three groups of plant viruses namely,

i. *Potyvirus*: *Sugarcane mosaic virus* (SCMV),
Maize dwarf mosaic virus (MDMV),
Johnson grass mosaic virus (JGMV) and
Sorghum mosaic virus (SrMV),

ii. *Tenuivirus*: *Maize stripe virus* or MStV (formerly MStpV), and

iii. *Rhabdovirus*: *Maize mosaic virus* (MMV) has been reported to naturally infect the sorghum crop in different countries. MStV, MMV, SCMV, MDMV, JGMV and red stripe disease (SRSD) are economically important on sorghum in India.

Maize stripe virus on Sorghum

17. *Maize stripe virus* (MStV-S)

» Virus causes wide range of symptoms on sorghum that include mosaic, mottling, yellowing, chlorotic streaks or stripes, reddening of leaves, necrotic spots, dwarfing/ stunting, delayed flowering, sterility and poor exertion of panicle depending on virus and stages of infection.

» The earliest symptoms of viral diseases under field conditions often resemble symptoms associated with herbicide, insecticide, fungicide damage or with genetic abnormalities.

18. *Maize mosaic virus* (MMV-S)

» The symptoms are characterized by fine discontinuous chlorotic streaks between veins on leaf.

» The lesions become necrotic as the disease progress.

» Infected plants become stunted in growth with short internodes.

» Early infected plant dies sooner or later without emergence of earhead.

» Plants infected at later stages may develop earhead with or without grain.

» It should be noted that there is chance of multiple virus infection in a plant under natural field conditions.

» The MStV-S and MMV-S are transmitted by insect vector *Peregrinus maidis*, the delphacid plant hopper on sorghum

19. Red stripe disease (SRSD)

» The disease is characterized by systemic symptoms of mosaic followed by necrotic red stripe and temperature dependent red leaf and general necrosis in sorghum.

Manangement of viral diseases

» Spraying metasystox 35 EC or methyl-S-demeton 35EC at 5 ml/ 10 lit.

» Application of carbofuran 3G at 8 kg/ ha inside the leaf whorls can also be done when the crop is at vegetative stage.

Minor diseases in India

» **Milo disease** (*Periconia* root rot) - *Periconia circinata* (M. Mangin) Sacc. & D. Sacc.

» **Pokkah Boeng** (Twisted top) - *Gibberella fujikuroi* (Sawada) Wollenw.

  Syn: *Fusarium moniliforme* var. *subglutinans* Wollenweb. & Reink.;
  G. fujikuroi var. *subglutinans* Edwards
  G. intermedia (Kuhlmann) Samuels, Nirenberg, & Seifert
  Syn: *F. proliferatum* (Matsushima) Nirenberg ex Gerlach & Nirenberg

» **Southern sclerotial rot**: *Sclerotium rolfsii* Sacc.

  Syn: *Athelia rolfsii* (Curzi) C.C. Tu & Kimbr.

» **Tar spot**: *Phyllachora sacchari* P. Henn.

» **Target leaf spot**: *Bipolaris sorghicola* (Lefebvre & Sherwin) Alcorn

  Syn: *B. cookei* (Sacc.) Shoemaker; *Helminthosporium cookei* Sacc.

Diseases of Sorghum

» **Zonate leaf spot and sheath blight**: *Gloeocercospora sorghi* Bain & Edgerton ex

» ***Ascochyta* blight**: *Ascochyta sorghi* Sacc.

» **Bacterial leaf blight**: *Acidovorax avenae* subsp. *avenae* (Manns) Willems *et al.*

» **Bacterial leaf spot**: *Pseudomonas syringae* pv. *syringae* van Hall

» **Bacterial top and stalk rot**: *Dickeya chrysanthemi* R. Samson *et al.*

Syn: *Erwinia chrysanthemi* Burkholder, McFadden, & Dimock

Chapter - 7

Diseases of Pearl Millet / Bajra / Cumbu - *Pennisetum glaucum* (L.) R.Br.

S. No.	Disease	Pathogen
1.	Graminicola downy mildew	*Sclerospora graminicola*
2.	Smut	*Tolyposporium penicillariae*
3.	Rust	*Puccinia pennisetti*
4.	Ergot	*Claviceps fusiformis*
5.	*Pyricularia* leaf spot (blast)	*Pyricularia grisea*

1. Graminicola downy mildew / Green ear

In India, this disease was first reported by Butler in 1907.

Symptoms

- » Infection is mainly systemic and symptoms appear on leaves and inflorescence.

- » The initial symptoms appear in seedlings at three to four leaf stages.

- » The affected leaves show patches of light green to light yellow colour on the upper surface and the corresponding lower surface bears dull white downy growth consisting of sporangiophores and sporangia.

Diseases of Pearl millet

» The yellow discolouration often turns to streaks along veins. As a result of infection young plants dry and die ultimately.

» Symptoms may appear first on the upper leaves of the main shoot or the main shoot may be symptom free and symptoms appear on tillers or on the lateral shoots.

» The inflorescence of infected plants gets completely or partially malformed with florets converted into leafy structures, giving the typical symptom of **green ear**.

» Infected leaves and inflorescences produce sporangia over a considerable period of time under humid conditions and necrosis begins. The dry necrotic tissues contain masses of oospores.

Chromista: *Sclerospora graminicola* (Sacc.) Schroet.

The mycelium is systemic, non septae and intercellular. Short, stout, hyaline sporangiophores arise through stomata and branch irregularly, with stalks bearing sporangia. Sporangia are hyaline, thin walled, elliptical and bear prominent papilla. Oospores are round in shape, surrounded by a smooth, thick and yellowish brown wall.

Disease cycle

The oospores remain viable in soil for 5 years or longer giving rise to the primary infection on seedlings. Secondary spread is through sporangia produced during rainy season. The dormant mycelium of the fungus is present in embryo of infected seeds.

Favourable conditions

» High humidity (90%), presence of water on the leaves favours the infection.

» Low temperature of 15-25°C favors the formation of sporangiophore and sporangia.

Management

» Deep ploughing to bury the oospores.

» Roguing out infected plants and adopt crop rotation.

» Grow resistant varieties like NPH 5, PHB 10, PHB 14, X 5, WCC 75, Co 6, Co 7, Cumbu Hybrid Co 9 and Co (Cu) 9.

» Seed treatment with metalaxyl at 6g/ kg.

>> Spray zineb 2.5 g/ lit or mancozeb 2 kg or metalaxyl + mancozeb at 1 kg/ ha or myazofamid 10 mg/ lit on 20th day after sowing in the field nearly at boot leaf stage.

2. Smut

Bajra smut occurs in Pakistan, Africa and India. It is prevalent in Tamil Nadu, Andhra Pradesh and Maharashtra. It generally causes limited damage to the crop; the heavy incidence in certain years may cause considerable loss in grain yield.

Symptoms

>> The pathogen infects few florets and transforms them into plump sori containing smut spores.

>> The sori are larger than normal healthy grains and when the sori mature they become dark brown releasing millions of black smut spore balls.

Fungus: *Tolyposporium penicillariae* Bref.

The fungus is mostly confined to the sorus. The sori contain spores in groups and are not easy to separate. Each spore is angular or round and light brown. Two other smuts of bajra, caused by *T. ehrenbergii* and *T. bullatum* Schroet, are minor in India.

Disease cycle

>> The pathogen survives as spore balls in the soil and serves as primary source of inoculum. Secondary spread is by air-borne conidia.

Favourable conditions

>> High relative humidity and successive cropping with pearlmillet.

Management

>> The damage caused by the fungus is negligible.

>> Removal and destruction of affected ear head will help in controlling the disease.

>> Seed treatments with raw cow milk or raw goat milk 50% dilution with water for 18 h with *Gliocladium virens* 6 g / kg and soil treatment with *G. virens* 10 g / m².

Diseases of Pearl millet 123

> Spray carboxin 0.15% at boot leaf stage two times at 10 days intervals.

3. Rust

Rust is a serious disease of bajra, occurring wherever the crop is grown. In India it is present in all the bajra tracts, causing considerable damage to some varieties, from the seedling stage to maturity.

Symptoms

> Symptoms first appear mostly on the distal half of the lamina.

> The leaf soon becomes covered by uredosori which appear more on the upper surface.

> The pustules may be formed on leaf sheath, stem and on peduncles.

> Later, telial formation takes place on leaf blade, leaf sheath and stem.

> While brownish uredia are exposed at maturity, the black telia remain covered by the epidermis for a longer duration.

Fungus: *Puccinia pennisetti* Zimm. (*Puccinia substriata* var. *penicilliariae*)

Uredospores are oval, elliptic, sparsely echinulated and pedicellate. Teliospores are dark brown in colour, two celled, cylindrical to club shaped, apex flattened, broad at top and tapering towards base. The fungus is macrocyclic producing uredial and telial stages on pearlmillet and aecial and pycnial stages on brinjal.

Disease cycle

The fungus has a long life cycle, producing the uredial and telial stages on bajra and the aecial and pycnidial stages on several species of *Solanum,* including brinjal (egg plant) (*Solanum melongena* L.). Air-borne uredospores are the primary sources. The uredial stages also occur on several species of *Pennisetum,* which helps in secondary spread of the pathogen.

Favourable conditions

> Closer spacing.

> Presence of abundant brinjal plants and other species of *Solanum viz., S. torvum, S. xanthocarpum* and *S. pubescents.*

124 Diseases of Field Crops and their Management

Management

» Grow resistant varieties like IP 537-B, P 2890, P 1577, P 1581 and P 2880.

» Spray with wettable sulphur 3 kg or mancozeb 2 kg/ ha.

» Spray with mancozeb 0.1% or captan 2.5 g/ lit or zineb 2.5 g/ lit thrice at 10-day interval starting at ear emergence.

4.   Ergot / Sugary disease

This disease is reported from many parts of Africa and India. In our country, it is on the increase during the past few years in Maharashtra and Karnataka, causing severe damage to the crop and poisoning the cattle which consume the ergot along with the straw.

Symptoms

» The symptom is seen by exudation of small droplets of light pinkish or brownish honey dew from the infected spikelets.

» Under severe infection many such spikelets exude plenty of honey dew which trickles along the earhead. This attracts several insects.

» In the later stages, the infected ovary turns into small dark brown sclerotium which projects out of the spikelet.

Fungi: *Claviceps fusiformis* Loveless 1967; *Claviceps microcephala* (Waller) Tul.

The pathogen produces septate mycelium which produces conidiophores and is closely arranged. Conidia are hyaline and one celled. The sclerotia are small (3-8mm × 0.3-15mm) and dark grey but white inside.

Disease cycle

Sclerotia are viable in soil for 6-8 months. The primary infection takes place by germinating sclerotia present in the soil. Secondary spread is by insects or airborne conidia. The role of collateral hosts like *Cenchrus ciliaris* and *C. setigerus* in perpetuation of fungus is significant. The fungus also infects other species of *Pennisetum.*

Management

» Adjust the sowing date so that the crop does not flower during September when

Diseases of Pearl millet

high rainfall and high relative humidity favour the disease spread.

» Immerse the seeds in common salt 10% solution and remove the floating sclerotia.

» Remove collateral hosts.

» Spray with carbendazim 500 g or mancozeb 2 kg or ziram 1 kg/ ha when 5-10% flowers have opened and again at 50% flowering stage.

» Spray ear-head with fungicides like zineb 2.5 g/ lit or ziram 0.1-0.15% or cuman-L 200 ppm or aureofungin or 8-hydroxyquinoline 500 ppm

5. *Pyricularia* leaf spot (blast)

Symptoms

» The disease appears as grayish, water-soaked lesions on foliage that enlarge and become necrotic, resulting in extensive chlorosis and premature drying of young leaves.

» Lesions are often surrounded by a chlorotic halo, which turns necrotic, giving the appearance of concentric rings.

» The lesions are usually confined to interveinal spaces on the foliage.

» Lesions grow and coalesce to cover large surface areas and cause necrosis of tissues.

» In case of a susceptible cultivar the entire foliage gives a burnt appearance.

» Severely infected plants produce no grain or few shrivelled grains in blasted florets.

Fungus: Teleomorph: *Magnaporthe grisea* (Herbert) Barr
Anamorph: *Pyricularia grisea* (Cooke) Sacc.

Asexual conidia are pyriform, hyaline, and mostly three-celled with a small appendage on the basal cell.

Disease cycle

The pathogen sporulates profusely in the lesions on foliage and the conidia can be easily dispersed by the wind and splashing rain. These spores can overwinter in stubble and can infect the next crop the following year.

126   Diseases of Field Crops and their Management

Favourable conditions

» Prevalence of high humidity (>90%) and moderate temperature (25-30°C) favors infection and disease development.

Management

» Grow resistant lines like ICMB 01333, ICMB 01777, ICMB 02111, ICMB 03999, ICMB 93222, ICMB 97222, ICMR 06222, ICMR 06444 and ICMR 07555

» Spray with carbendazim 2 g/ lit or tricyclozole 2 g/ lit at pre-flowering stage.

Minor diseases

» **Grain mould:** Fungal complex: Grains covered with white, pink or black moulds.

» **Zonate leaf spot:** *Gloeocercospora* sp.

» **Banded leaf spot:** *Rhizoctonia* spp.

» **Leaf blast:** *Pyricularia seteriae* Nishikado

» **Smut:** *Tolyposporium pennicillariae; Tolyposporium senegalense; Tilletia ajrekari*

» **Eyespot:** *Dreshlera sacchari; Drechslera (Helminthosporium) australiense*

» **Leaf spot:** *Curvularia lunata; Curvularia penniseti* (Mitra) Boedijn

» **Chacoal rot:** *Macrophomina phaseolina* (Tassi) Goid.

» **Southern blight:** *Sclerotium rolfsii* Sacc.

» **Phyllosticta leaf blight:** *Phyllosticta penicillariae* Speg., (1914)

» **Top rot / Twisted top / Pokah Boeng:** *Fusarium moniliforme* Sheld.

» **Leaf spot:** *Xanthomonas penniseti* (*Xanthomonas campestris* pv. *penniseti*)

» **Leaf blotch:** *Xanthomonas annamalaiensis* (*Xanthomonas campestris* pv. *annamalaiensis*)

» **Reniform nematode:** *Rotylenchus reniformis*

Integrated management strategies for major pest and diseases of pearl millet

Seed treatment with metalaxyl at 6g/ kg + Seed treatment with imidacloprid at 5g/ kg + Removal of downy mildew infected plants up to 45 days of sowing + Spraying of mancozeb at 1000 g/ ha + Spraying of NSKE 5% at 50% flowering against downy mildew, rust and shoot fly.

Chapter - 8

Diseases of Finger millet / Ragi - *Elesine coracana* Gaertn.

S. No.	Disease	Pathogen
1.	Blast	*Magnaporthe grisea*
2.	Seedling blight / Leaf blight	*Helminthosporium nodulosum*
3.	Smut	*Ustilago eleusine*
4.	Downy mildew	*Sclerophthora macrospora*
5.	Foot Rot	*Sclerotium rolfsii*
6.	Banded blight	*Rhizoctonia solani*
7.	Sheath blight	*Marasmius candidus*
8.	Bacterial leaf spot	*Xanthomonas eleusineae*
9.	Bacterial blight	*Xanthomonas coracanae*
10.	Bacterial leaf stripe	*Pseudomonas eleusineae*
11.	Ragi mottle streak disease	*Ragi mottle streak virus*
12.	Ragi severe mosaic disease	*Sugarcane mosaic virus*
13.	Ragi streak disease	*Eleusine* strain of *Maize streak virus*

1. Blast

Among the several diseases that afflict finger millet, blast (*Pyricularia grisea*) is not only widely distributed in almost all the finger millet growing regions of the world, but also it is the most destructive disease. The disease was reported for the first time in India, from Tanjore delta of Tamil Nadu by Mc Rae (1920).

Symptoms

» The disease occurs at all the stages of plant growth *viz.,* germlings to earhead and even on seed.

» In nursery or in the direct sown crop, the death of the germ lings is common, especially if the seeds used for sowing are infected.

» The lesions are spindle shaped and are of different size.

» In the beginning, the spots have yellow margin with grayish green centre.

» Under humid conditions, an olive grey overgrowth of fungus can be seen on the centre of the spot.

» Later the centre become whitish grey and disintegrates.

» The lesions on the seedlings are about 0.3-0.5 cm in breath and 1-2 cm in length.

» Stem infection causes blackening of the nodal region.

» The neck region turns black and shrinks.

» Infection may also occur at the basal portions of the panicle branches including the fingers.

» The affected portions turn brown and ears become chaffy and only few shrivelled grains are formed.

Fungus: Teleomorph: *Magnaporthe grisea* (Herbert) Barr
Anamorph: *Pyricularia grisea* (Cke.) Sacc.

The hyphae are intracellular or inter cellular, hyaline and septate and turns to brown when become old. Numerous conidiophores and conidia are formed in the middle portion of the lesions. Conidiophores are slender, thin walled, emerging singly or in groups, unbranched, geniculate and pale brown in colour. Conidia are thin walled, sub-pyriform, hyaline 1-2 septate, mostly 3 celled with a prominent hilum.

Diseases of Finger millet

Favourable conditions

The more favourable for blast development with average minimum and maximum atmospheric temperatures around 20°C and 30°C respectively and a relative humidity of more than 80% *Pyricularia* from ragi also infected barley and wheat but not maize, oats, rice, sorghum, *Panicum miliaceum,* etc.,

Rice and ragi pathogens were not cross inoculable. *Brachiaria mutica* and *Panicum repens* to be collateral hosts of ragi blast. In addition, the pathogen readily infects *Setaria italica, Eleusine indica, Lolium multiforum, Festuca elatior, Phalaris arundinaceae, Anthoxanthum odoratum, Dactyloctaenium aegiptium,* wheat, barley, oats and maize etc.,

Management

» Use of blast resistant varieties like Co 10, Co 11, Co 12, Co 13, GPU 28, GPU 26, and GPU 48 coupled with carbendazim as seed treatment at 2 g/ kg.

» Soil application of potassium 30 kg combined with KCl 1% foliar spray gave the maximum disease reduction and the highest yields.

» Seed treatment with captan 4 g or carbendazim 2 g/ kg or *Pseudomonas fluorescens* at 10 g/ kg.

» Spray any one of the fungicides edifenphos 500 ml or carbendazim 500 g or iprobenphos at 500 ml/ ha or Zineb 1.5 kg/ ha, first spray immediately after noticing the symptoms, second and third sprays at flowering stage at 15 days interval.

» Foliar spray with Aureofungin-sol 100 ppm at 50% earhead emergence followed by a second spray with mancozeb 1000 g/ ha or *Pseudomonas fluorescens* at 0.2% 10 days later.

2. Seedling blight / Leaf blight

Seedling blight or leaf blight of finger millet is next only to blast both in severity and distribution. The disease generally becomes severe during the advance stage of plant maturity, as it happens to be a low sugar disease.

Symptoms

» The pathogen attacks all the parts of the plants including roots, base of the plants, culms, leaf sheath, leaf blade, neck of the panicle and the fingers.

130 Diseases of Field Crops and their Management

» Both pre-and post-emergence rot may be seen.

» On young leaves the disease appears as minute, light brown oval spots.

» The affected leaves wither prematurely and seedlings may be killed.

» The fungus affects the base of the plants and cause root rot and foot rot. In grown up plants, spots are oblong and dark brown.

» The spots on the leaf sheath and culms are irregular and are generally found on the junction of blade and sheath.

» Infection on the neck causes discoloration and sooty growth in the inflorescences.

Fungus: Teleomorph: *Cochliobolus nodulosum* Luttr.,
Anamorph: *Helminthosporium nodulosum* Berk. and Curt.

Hypha of the fungus is light brown coloured and septate. Conidiophores are long, septate, dark brown in colour, often branched and geniculate. Conidia are straight ovoid, pale to dark golden brown, 5-7 pseudoseptate. *C. nodulosum* produces spherical perithecia and asci contain 1 to 8 ascospores.

Management

» Seed treatment with captan 0.25%.

» Spray mancozeb 0.2%.

3. Smut

Kulkarni (1922) recorded this disease from Malkapur in 1918 in the then princely state of Kolhapur, described the pathogen and identified it as *Ustilago eleusine.*

Symptoms

» The disease generally manifests itself few days after flowering.

» The smutted grains can be seen scattered randomly in the ear head.

» The affected ovaries get converted into greenish gall like bodies which are several times bigger in size than the normal grains.

» In the initial stages, greenish swollen grains 2-3 mm in diameter are evident which

Diseases of Finger millet

project beyond the glumes.

» Later the infected grains become swollen and can reach a diameter of about 16 mm.

» The greenish outer tunica of the sorus gradually turns pinkish-green and finally turns to dirty black on drying.

» Some times the affected grains are single or grouped in patches of varying sizes and are frequently confined to one side or towards the base or apex of the head and show signs of rupturing at several places.

» Spores are formed in the cavities found in the sori, mixed with a gelatinous matrix but later the mass becomes pulvurent.

Fungi: *Ustilago eleusine* Kulk; *Melanopsichium eleusinis* (Kulk) Mundk and Thirum

The pathogen is neither systemic nor externally seed borne. Floral infection is by air-borne spores.

Management

» Spray fosetyl-Al 0.2% and carboxin 0.3%.

» Spraying first with captafol 0.2% at panicle initiation followed by a spray with mancozeb 0.2% at flowering stage was recommented.

4. Downy mildew

Downy mildew is one of the highly destructive diseases of finger millet wherever it occurs. Incidentally its occurrence is only sporadic. This disease was reported for the first time in India by Venkatarayan (1946) on ragi from the erstwhile Mysore state.

Symptoms

» Downy mildew affected plants are generally stunted with shortened internodes and profuse tillering.

» The leaves are crowded giving thereby a bushy appearance to infected plants.

» If infected in seedling stage, the seedlings may get killed.

» Often, pale yellow translucent spots are seen on leaves of affected plants.

» The most striking feature of the disease however, is partial or complete proliferation of the spikelet into leafy structures which often result into a bush like structure. Proliferation of floral organ is common.

Chromista: *Sclerophthora macrospora* (Sacc.) Thirllm, Shaw and Naras
Syn: *Sclerospora macrospora* Sacc

Mycelium is coenocytic, hyaline. Sporangiophores are sympodially and successively branched, and haploid in nature. Sporangia are lemon shaped, borne singly at the apices of sporangiophore. Oospore germination is indirect by the formation of a big lemon-shaped sporangium which liberates 24-48 zoospores.

Management

» Seed treatment with metalaxyl 0.2% or mancozeb + metalaxyl 0.2%.

5. Foot rot / Wilt

Foot rot in fingermillet is caused by a non-specialized pathogen *Selerotium rolfsii.* Coleman (1920) was probably the first in India to record occurrence of *S. rolfsii* from the princely state of Mysore. Sundararaman (1933) reported S. *rolfsii* from wilted ragi in former Madras Presidency. Further, it was recorded at Coimbatore.

Symptoms

» The basal portion, immediately above the ground, initially appears water soaked due to the infection by the pathogen.

» Later on this portion turns brown and subsequently dark brown with a concomitant shrinking of the stem in the affected region.

» The infected plants become pale, chlorotic and stunted.

» White cottony mycelial growth occurs profusely in this area. On the surface of the lesions, small spherical, dark coloured sclerotia are formed.

» Small roundish white velvety grain like structures starts appearing in the fungal matrix.

» These sclerotial bodies grow, attain the size of mustard seeds, and turn brown.

Diseases of Finger millet 133

> Meanwhile the leaves loose their luster, droop and dry.

> Ultimately the plant dries up prematurely, such affected plants are seen randomly in the field.

Fungus: Anamorph: *Sclerotium rolfsii* Sacc. (Teleomorph: *Pellicularia rolfsii*)

The mycelium of the fungus is septate and white to tan coloured. Sclerotia are minute, mustard seed like structures and black in colour.

Management

> Soil drenching with carboxin at 10 kg/ ha.

6.   Banded blight

Lulu Das and Girija (1989) for the first time reported sheath blight of ragi from Vellayani in Kerala, India, where it occurred in a severe form.

Symptoms

> The disease is characterized by oval to irregular light grey to dark brown lesions on the lower leaf sheath.

> The central portions of the lesions subsequently turn white to straw with narrow, reddish, brown border.

> Such spots, at later stages, are distributed irregularly on leaf lamina.

> The disease symptom is characterized by a series of copper or brown colour bands across the leaves giving a very characteristic banded appearance.

> Later on, the leaves dry up and plants appear blighted.

> On peduncles, fingers and glumes irregular to oval, dark brown to purplish brown necrotic lesions are formed.

> Early infection on peduncle or near finger base is somewhat similar to neck rot resulting in poor grain filling.

> If the sheath is infected before peduncle emergence, then the fingers are disorganised and reduced in size.

134   Diseases of Field Crops and their Management

» Infected glumes produce smaller and shriveled grains.

» Thus, the symptoms produced on every part of the plant give a very characteristic banded appearance, due to which the disease has been named as banded blight.

Fungus: *Rhizoctonia solani* Kuhn (Sclerotial stage);
           Thanatephorus cucumeris (Fr.) Donk. (Basidial stage)

The mycelial growth along with white to brown sclerotia can be observed on and around the lesions.

Favourable conditions

» A temperature of 28-30 °C and a relative humidity of 70%.

Management

» Deep summer ploughing

» Soil drenching with FYM 50 kg + *Pseudomonas fluorescens* 2.5 kg.

7.   Sheath blight: *Marasmius candidus* Bolt.

Sheath blight of ragi caused by a mushroom was reported for the first time from Coimbatore, Tamil Nadu, India in 1974 (Parambaramani *et al.*, 1975).

Symptoms

» The first symptom of the disease is the appearance of characteristic circular to elliptic, necrotic patches on the sheaths about 5 to 15 cm above the ground level.

» As the disease progresses, the sheaths get stuck or bound together, with the mycelium of the fungus to the stem eventually leading to wilting and death of the plants.

» The diseased plants in the field could be easily distinguished by the discoloured or dried up outer leaves.

» On lower sheaths of the dead plants small sporophores of the mushroom are noticed.

Diseases of Finger millet   135

8. Bacterial leaf spot: *Xanthomonas eleusineae* Rangaswami, Prasad, Eswaran

Rangaswami *et al* (1961) reported a bacterial disease from Tamil Nadu, which they identified as *Xanthomonas eleusinae*. Later on Patel and Thirumalachar (1965) reported yet another species of *Xanthomonas* from Gujarat.

Symptoms

» Spots are seen on both upper and lower surface of the leaf blade.

» They are linear and spread along the veins. The spot measure 2 - 4 mm long, but often extends up to one inch or more.

» In the beginning spots are light yellowish brown, but soon become dark brown.

» In advanced stage, the leaf splits along the streak, giving a shredded appearance.

9. Bacterial blight: *Xanthomonas coracanae* Desai, Thirumalachar and Patel.

Symptoms

» The plants are susceptible to infection at all stages of growth.

» If infection takes places during early stages of growth, the plants become yellow and show premature wilting.

» Infection first appears as water soaked, translucent, linear, pale yellow to dark greenish-brown streaks, 5-10 mm long, and parallel to the midrib of the lamina.

» The hyaline streak later develops into a broad yellowish lesion measuring 3-4 cm and turns brown.

» In severe infection, particularly in the early stages, the entire leaf turns brown and withers away.

10. Bacterial leaf stripe: *Pseudomonas eleusineae*

Symptoms

» The common symptom of the disease is brown coloration of the leaf sheath especially from base upwards.

» The affected portion of the lamina invariably involves the midrib and appears straw colored.

» This symptom spreads to about three fourths the lamina and then abruptly stops or in some cases reaches the leaf tip.

» Occasionally the strips of infected areas are seen to proceed along the margin of the lamina.

» The bacteria are readily detected in the phloem vessels.

» Infected plants can be recognized from distance by characteristic drooping of the leaves.

11. Ragi mottle streak disease

During mid: 1966's there was a severe disease problem in epidemic proportions althrough Southern Karnataka. Early this disease and thought it was due to combined effect of a Virus and *Helminthosporium* sp. It was *Rhabdovirus* transmitted by the fulgorid insect as *Sogatella longifuraifera* (E&I) and reported two more leaf hoppers *viz.*, *Cicadulina bipunctella* (Mats) and *Cicadulina chinai* (Ghauri) which were able to transmit.

Symptoms

» The infected plants exhibit regular dark-green areas all along the leaf veins when the plants are 4-6 weeks old.

» Other symptoms on leaf include chlorosis and streaking.

» In some cases occasional yellowing to almost albino symptoms are also observed.

» However, in the lower leaves the symptoms are of a mottle type in the form of white specks and the affected plants are generally stunted bearing small ear heads.

Virus: *Ragi mottle streak virus*

Favourable conditions

The disease is high in April-May sown crops due to high population of vectors *viz.*, *Cicadulina bipunctella* and *C. cinai.*

Management

» Rogue out the affected plants.

» Spray insecticides like methyl dematon 25 EC 500 ml/ ha at 20 days intervals.

Diseases of Finger millet 137

12. Ragi severe mosaic disease

The ragi crop in Southern Karnataka and the border districts of Andhra Pradesh was affected by a severe mosaic in *kharif* 1966. In certain areas like Hiriyur Taluk in Chitradurga district and Devanahally taluk of Bangalore district are highly prone to this disease.

Virus: *Sugarcane mosaic virus*; Transmitted by *Longiunguis sacchari* (Aphid) as vector.

Symptoms

» The mosaic pattern is very clear and pronounced on young leaves.

» Diseased plants remain stunted and the ear heads of severely affected plants are malformed.

» In addition to the above symptoms, observed that the affected plants to be pale yellow due to severe chlorosis and in severe cases they became brownish-white.

» Thus, the entire field appears yellow and can be readily distinguished from non-infected stands from a distance.

» Stunted plants do not recover, develop roots at nodes, generally do not produce tillers and if produced remain mostly sterile.

» In severe case severe mottling, chlorotic streak, general chlorosis and yellowing of leaves.

» They also observed profuse lateral shoots and aerial adventitious roots, stunting of plants, fewer flower and poor seed formation.

13. Ragi streak disease

Virus: *Eleusine* strain of *Maize streak virus*; Transmitted by *Cicadulina chinai.*

» Symptoms appear on young unfolding leaves as pale specks or stripes of different size.

» They subsequently expand larger areas resulting in chlorotic bands running almost the entire length of the leaf parallel to the midrib.

» These bands are occasionally interrupted by dark green areas.

» The new emerging leaves of both the main shoot and the tillers show number of well defined chlorotic streaks having almost uniform width running parallel to the midrib throughout the length of the leaf lamina.

» The infected plants in the field produce comparatively more number of tillers and bear yellowish sickly earheads, often bearing few shrivelled seeds.

» The plants infected very early in the crop growth stage die before they bloom.

Minor diseases

» ***Cercospora* leaf spot**: *Cercospora eleusinis*

» **Damping-off**: *Pythium aphanidermatum*

» **Leaf spot** - *Curvularia lunata* (Wakker) Boedljin

» **Ozonium wilt** - *Ozonium texanum* var. *parasiticum*

» **Rust** – *Uromyces erogrostidis* Tracy

» **Witch weed (Phanerogamic parasite)** - *Striga asiatica* (L.) Kuntze and *Striga densiflora* (Benth.) Benth.

Chapter - 9

Diseases of Foxtail millet / Thenai Italian millet / Kakun – *Setaria italica* (L.) P. Beauvois.

S. No.	Disease	Pathogen
1.	Blast	*Pyricularia setariae*
2.	Smut	*Ustilago crameri*
3.	Rust	*Uromyces setariae-italicae*
4.	Downy mildew	*Sclerospora graminicola*
5.	Leaf spot	*Cochliobolus setariae*

1. Blast

Symptoms

»   The spots are seen on the leaf blade.

»   They are circular with light centre and are surrounded by a dark brown margin.

»   The spots are small and scattered. When the disease appears in severe form the leaves wither and dry up.

»   Neck infection is very rare.

Fungus: *Pyricularia setariae* Nishikado.

The conidiophores emerge through epidermal cells or through stomata. Several conidia

140   Diseases of Field Crops and their Management

are formed one after another from each conidiophore. They are sub-hyaline, three celled and obpyriform. Thick walled, olive-brown and globose chlamydospores are also developed at the tips of the germ tube.

Disease cycle

- » The primary source of infection is through seed-borne conidia and to some extent soil-borne.

- » The secondary spread is through air-borne conidia which are produced on ragi, bajra, wheat and *Dectylotacnium aegyptium*.

Favourable conditions

- » The optimum temperature is 30°C.

- » High relative humidity 90%, low night temperature and cloudy weather.

Management

- » Grow resistant varieties like SR 118, SR 102, ISc 709, 701, 703, 710, 201, JNSc 33, 56, RS 179 and ST 5307.

- » Seed treatment with captan 4 g or carbendazim 2 g/ kg.

- » Spray the crop with iprobenphos or edifenphos 500 ml/ ha.

- » Top dressing of nitrogen has to be taken up after the fungicidal spray.

2.   Leaf spot / Leaf blotch

Symptoms

- » Leaf spots are brown in colour and small.

- » Some times lesions also appear as blotches and the rotting of the secondary roots may also occur.

Fungus: *Cochliobolus setariae* (Ito and Kuribayashi) Drechsler ex Dastur.
         Syn: *Helminthosporium setariae* Sawada

The conidiophores are simple, erect, cylindrical, brown, slightly swollon at the base

Diseases of Foxtail millet

and geniculate at the apex. Conidia are ellipsoid, straight or slightly curved and pale to moderately dark brown.

Favourable conditions

- » Optimum temperature for growth and sporulation is 30°C
- » Externally seed borne.

Management

- » Seed treatment with captan at 4 g/ kg.

3. Smut

Symptoms

- » The fungus grows systemically inside the host and expresses the symptom at the time of flowering.
- » The sori are seen in the flowers and the basal parts of the palea.
- » The fungus affects most of the grains in an ear but sometimes the terminal portion of ear may escape.
- » The sori are pale grey in colour and measures 2 to 4 mm in diameter. When the crop matures the sori rupture and liberate dark powdery mass of spores.

Fungus: *Ustilago crameri* Korn.

- » The chlamydospores are dark brown in mass but lighter singly, irregular or angular in shape and smooth walled.
- » The chlamydospores are inter-calary in hyphal strands.

Disease cycle

- » The fungus is externally seed-borne and secondary spread by air-borne chlamydospores. Soil borne infection has also been observed.

Favourable conditions

» Low temperature and high relative humidity are favourable for rapid development and spread of the disease.

Management

» Seed treatment with captan at 4 g/ kg.

» Steeping the seeds in copper sulphate 2% or formalin 0.5% for about 30 minutes.

4. Rust

Symptoms

» Numerous minute, brown uredosori appear on both surface of the leaf and are covered by the epidermis for very long time.

» The pustules are small, oblong and cinnamon brown in colour.

» The telia are smaller but covered by epidermis for quite a longer period and are grayish black in colour.

» Severe incidence of disease reduces the yield.

Fungus: *Uromyces setariae italicae*

The uredospores are round, spiny, yellowish brown with 3 or 4 germpores. The teliospores are one celled, smooth, oblong globose and thick walled especially at the apex.

Disease cycle

» The fungus can also attack other species of *Setaria viz., S. glauca, S. viridis* and *S. verticillata*. The air-borne uredospores cause primary infection.

Management

» Removal of collateral hosts

» Spray mancozeb 2.5 g/ lit.

Diseases of Foxtail millet 143

5. Downy mildew / Green ear

Symptoms

- » Primary infection causes chlorosis of the plant and the leaves turn whitish.

- » The terminal spindle fails to unroll, becomes chlorotic and later turns brown and gets shredded.

- » Whitish bloom of sporangiophores and sporangia develop on the surface of the affected leaves under humid conditions.

- » The affected plants rarely come to flowering. If the infection is mild, the plants may develop ears but the floral parts are proliferated into green leafy structures called green ear.

Fungus: *Sclerospora graminicola* (Sacc.) Schroet.

The fungus is an obligate parasite. Primary infection is mainly from soil-or seed borne oospores. The sporangiophores are quite and branch heavily. The sporangia are hyaline, broadly fusiform or ovate in shape. Oospores are also produced in the infected host tissue. They are spherical with smooth well and dark brown in colour.

Disease cycle

Primary infection is mainly from soil-borne oospores or from oospores on the grains. The oospores are able to survive upto 8-10 years.

Favourable conditions

- » Rainy weather, low temperature 15-25°C, high humidity 90% and high soil moisture.

Management

- » Seed treatment with captan 4 g/ kg.

Minor diseases

- » **Bacterial blight:** *Pseudomonas alboprecepitans* Rosen

- » **Bacterial brown stripe:** *Pseudomonas setariae* (Okabe) Savul.

- » Syn: *Bacterium setariae*
- » **Bacterial leaf spot:** *Xanthomonas indica*
- » **Udbatta diseases**: *Balansia oryzae*
- » **Kernal smut**: *Ustilago crameri*

Chapter - 10

Diseases of Indian Barnyard millet / Kuthiraivali *Echinochloa frumentacea* Link

S. No.	Disease	Pathogen
1.	Blast	*Magnaporthe grisea*
2.	Shoot and head smut	*Ustilago crusgalli*
3.	Grain smut	*Ustilago panicifrumentacei*
4.	Downy mildew	*Sclerophthora macrospora*
5.	Leaf spot / blight	*Helminthosporium crusgalli*

1. Downy mildew

Symptoms

» The initial symptoms appeared as a light yellow bands on the leaves which later turn into pale white in color.

» Later the leaves start drying and in severe case affected plants produced chaffy grains.

Fungus: *Sclerophthora macrospora* (Sacc.) Thirum.

Management

» Remove the diseased plants.

2. Shoot and head smut

Symptoms

- » The infected inflorescence is deformed and destroyed.
- » In addition, the smut also produces gall-like swellings on the stem, the nodes of young shoots and in the axils of the older leaves.
- » Sometimes, twisted, deformed clusters of leafy shoots with aborted ears may also develop.
- » The gall-like swellings are covered by a hairy rough membrane of host tissue and may be up to 12 mm in diameter.
- » The pathogen infects only ovaries.
- » The affected ovary is transformed into hairy, grey sac, round but does not increase in size than the normal grain.

Fungus: *Ustilago crusgalli* Tracy & Earle
The smut spores are mikado-brown, spherical and echinulated.

Favourable conditions

- » Externally seed-borne.
- » Temperature ranges with in 20 to 25°C

Management

- » Grow resistant genotypes like PRB 402, TNAU 92, VL 216 and PRJ 1.
- » Rouging of infected plants from the field is reducing the spread of disease.
- » Seed treatment with captan 4 g/ kg.

3. Grain smut

Symptoms

- » The grains in the ear are replaced by smut sori which contain black mass of powdery spores.

Diseases of Indian Barnyard millet

» The affected seeds enlarge two to three times of their normal size and their surface becomes hairy.

» Infected grains in case of *Ustilago panicifrumentacei* is larger than healthy grains.

Fungi: *Ustilago panicifrumentacei* Bref.; *Ustilago paradoxa* Syd. & But.

Management

» Grow resistant genotypes like PRB 402, TNAU 92, VL 216 and PRJ 1.

» Seed treatment with carbendazim or thiram 2 g /kg.

4. Leaf spot / blight

Symptoms

» The disease appears as isolated, dark brown, scattered and spindle shaped spots, measuring 0.5-3.0 mm × 0.25-1.5 mm, on flag leaves.

» Afterwards, several such spots coalesce and cover the entire leaf, which becomes grey and dried up.

» The spots are dark brown to grey in colour and are surrounded by yellow halo. Just after the appearance of the lesions, dark points are visible in the centre.

» Under humid conditions fungal growth is visible on these spots. In severe form the leaves show brightening.

» Similar spots can also be seen on the leaf sheath.

» The disease is most common under humid conditions.

Fungi: *Helminthosporium crusgalli* Nisikado and Miyake; *Helminthosporium frumentacei* Mitra; *Helminthosporium setariae*; *Helminthosporium monoceros* Drechsler

The disease is most common under humid conditions.

Management

» Spray copper sulphate 0.2%

5. Blast

Symptoms

- » The symptoms in form of spindle to circular shaped spots and of different sizes appear on the young seedlings in the field.

- » In the beginning the isolated spots have yellowish margin and grayish centre.

- » Later, the centres become ash coloured.

- » Under humid conditions, an olive-grey overgrowth of fungus having conidiophores and conidia develops at the centre of the spots.

- » At the later stage of infection the spots enlarge and coalesce.

Fungus: *Magnaporthe grisea* (Herbert) Barr

Favourable conditions

- » Temperature range of 25-30°C, humidity of 90% and above, cloudy days with discontinuous rainfall.

Management

- » Seed treatment with carbendazim 2 g/ kg.

Minor diseases

- » **Rust**: *Puccinia* sp.

- » **Ergot**: *Claviceps microcephala*

Chapter - 11

Diseases of Kodo millet / Varagu / Kodo - *Paspalum scrobiculatum* L.

S. No.	Disease	Pathogen
1.	Head smut	*Sorosporium paspali-thunbergii*
2.	Rust	*Puccinia substriata*
3.	Ergot	*Claviceps paspali*
4.	Udbatta	*Ephelis oryzae*
5.	Bacterial leaf streak	*Xanthomonas* sp
6.	Witch weed	*Striga* sp.

1.   Head smut

Symptoms

- »   The entire panicle is transformed into a long sorus and cream coloured thin membrane covers the sorus.

- »   In some cases it is enclosed in the flag leaf and may not emerge fully. The membrane bursts open and exposes the black mass of spores.

Fungus: *Sorosporium paspali-thunbergii* McAlp.

» Head smut of kodo millet is both seed and soil borne

» Spores are globose to angular and dark brown with a thick smooth epispore.

Disease cycle

» Externally seed-borne.

» The spores stick to surface of the grains and infect the next crop.

Management

» Grow resistant genotypes like GPLM 78, GPLM 96, GPLM 176, GPLM 322, GPLM 364, GPLM 621, GPLM 641, GPLM 679 and GPLM 720.

» Seed treatment with captan at 4 g/ kg or copper sulphate 1.5% or copper carbonate 6 g/ kg.

2. Rust

Symptoms

» The erumpent, oval, brown uredia are formed on the upper surface of the leaf blade and on the leaf sheath.

» The brown coloured telia are formed on the under surface of the leaf blade and on the leaf sheath

» The fungus produces cinnamon brown (uredia) to chocolate brown (telia) scattered pustules on the both surface of the leaves.

» The alternate host and pycnidial and aecial stages are not known.

Fungus: *Puccinia substriata* Ellis & Barth. (Syn: *Uredo paspali-scrobiculati* Sid.)

Management

» Spray mancozeb at 2 g/ lit.

3. Ergot / Sugary disease

Symptoms

» The honey-dew or conidial stage and the sclerotial stages of the fungus are known.

Diseases of Kodo millet

» The fungus infects forage grasses including *Paspalam dilatatum.*

» As the panicle matures, dark grey sclerotia or ergot replace the kernel which is the most conspicuous symptom of the disease.

Fungus: *Claviceps paspali* Stev. & Hall.

Management

» Clean cultivation

4. Udbatta disease

Symptoms

» The affected panicles are transformed into a compact agarbatti like shape, hence the name "Udbatta".

Fungus: *Ephelis oryzae* Syd. (Teleomorph: *Balansia oryzae sativae* Hoshioka)

Management

» Grow resistant genotypes like IPS 45, IPS 196, IPS 342, IPS 365, IPS 368, IPS 387, IPS 140 and Niwas 1.

» Removal and burning of affected panicles.

» Keeping the bunds free from weeds that serve as collateral hosts

» Pre-sowing seed treatment with carbendazim at 4 g/ kg.

5. Bacterial leaf streak

Symptoms

» The disease manifests itself as pale yellow streaks measuring 0.5 to 1.0 mm running parallel to the veins of leaf.

» Later, streaks enlarge to 1.0-1.5 mm × 3-4 cm lesions, which ultimately turn brown.

» In severe infection, the entire leaf withers away.

» The leaves may be shredded along the length.

152   Diseases of Field Crops and their Management

» Streaks may also be formed on shoots and peduncle of panicles.

Bacterium: *Xanthomonas* sp

Management

» Grow resistant varieties like JK 41, JK 62, Co 2, Co 3, T 1 and IPS 14

» Spray streptomycin sulphate 300 ppm.

6. Witch weed (Phanerogamic partial root parasite)

» The infestation of Striga species appears in the field after emergence of Striga plants from the soil.

» The under-ground portion of Striga plants remain attached to the roots of host plant by houstoria, from which the parasite absorbs water and nutrients.

» The attacked plants are stunted with poor aerial growth and bear lanky panicles. If the infestation occurs in early stage, the plants may dry up before flowering.

Parasite: *Striga* sp. (Witch weed)

Management

» Grow resistant varieties like JK 41, GPUK 3 and GPUK 5

» Pre flowering hand pulling of parasites.

Minor diseases

» **Leaf spot**: *Drechslera* state of *Trichometasphaeria holmii* Luttrell = *D. holmii* (Luttrell)

» Subram. & Jain (*Helminthosporium holmii*)

» **Leaf spot**: *Drechslera victoriae* (Meehan & Murphy) Subram. & Jain.

» **Leaf spot**: *Dimmerosporium* (=*Asterina*) *erysiphoides* Ell. & Ev.

» **Leaf blight**: *Drechslera* state of *Trichometasphaeria turcica* Luttrell (*Helminthosporium turcicum* Pass.).

Diseases of Kodo millet

» **Leaf blight**: *Alternaria alternata* (Fr.) Keissler.

» **Root rot**: *Macrophomina phaseolina* (Tassi) Goid

» **Sheath blight**: *Rhizoctonia solani* Kuhn

» **Leaf rust**: *Uredo paspali-scrobiculati* Syd.

» **Inflorescens infection**: *Ustilago rabenhorstiana* Kuehn.

Chapter - 12

Diseases of Proso millet / Panivaragu / Common millet - *Panicum miliaceum* L.

S. No.	Disease	Pathogen
1.	Head smut	*Sporisorium destruens*
2.	Grain smut	*Sphacelotheca sorghi*
3.	Rust	*Uromyces linearis*
4.	Downy mildew	*Sclerospora graminicola*
5.	Leaf spot	*Bipolaris panici-miliacei*
6.	Sheath rot	*Rhizoctonia solani*
7.	Udbatta	*Ephelis oryzae*
8.	Bacterial stripe	*Pseudomonas avenae*

1. Rust

Symptoms

» Numerous, narrow, minute, brown pustules arranged in linear rows appear on the upper surface of the leaves.

Fungus: *Uromyces linearis* Berk. & Br.

Uredia are erumpent and brown in colour. The fungus also attacks the other hosts like

Diseases of Proso millet

Panicum ripens and *P. antidotale*. The fungus is spread through air-borne uredospores.

2. Head smut

Symptoms

- » Infected panicle becomes elongated and thickened.
- » The entire inflorescence is converted into a smutted gall.
- » Further, the galls are converted into greyish white false membrane.

Fungus: *Sporisorium destruens* (Schlechtend) K. Vanky (Syn: *Sphacelotheca destruens*).

It is an externally seed borne fungal disease.

Management

- » Seed treatment with carboxin 0.2%.

3. Grain smut

Symptoms

- » Most of the grains are transformed into white grayish sacs (smut sori).
- » The sori are slightly pointed to oval and filled with black powder (chlaymydospores).

Fungus: *Sphacelotheca sorghi* (*Ustilago crameri*)

Management

- » Crop rotation for 2-3 years
- » Seed treatment with carboxin 0.2%.

4. Downy mildew

Symptoms

- » Ashy mycelial growth was observed on the lower surface of the leaves, followed by leaf shredding.

Chromista: *Sclerospora graminicola* (Sacc.) Schroet.

Management

» Seed treatment with metalaxyl 0.2%.

5. Leaf spot

Symptoms

» The fungus which attacks the rice also attacks this crop and produce brown rectangular spots.

Fungus: *Bipolaris panici-miliacei* (Syn: *Helminthosporium panici-miliacei*)

Management

» Grow resistant varieties like RAUM 7

6. Sheath rot

Symptoms

» The disease starts at the tillering stage of the crop and continues in the succeeding stages.

» First symptoms appear as greyish green lesions on the leaf sheath between soil level and leaf blade.

» The lesions are ellipsoid or ovoid, 2-3 cm long that become greyish white with brown margins.

» The disease spreads upwards causing blighting of leaf sheath and leaf blades.

Fungus: *Rhizoctonia solani* Kuhn

Management

» Crop rotation for 2-3 years

» Seed treatment with carboxin 0.2%.

Diseases of Proso millet

7. Udbatta

Symptoms

> The affected panicles are transformed into a compact agarbatti like shape, hence the name "Udbatta".

Fungus: *Ephelis oryzae* Syd. (Teleomorph: *Balansia oryzae-sativae* Hoshioka).

Management

> Crop rotation for 2-3 years

8. Bacterial stripe

Symptoms

> Narrow, brown, water soaked streaks extend from the blades of the leaves down to the sheaths and also on culms, where many streaks coalesce.

> The tissue turns brown and translucent.

> Abundant exudates are evident in the form of thin, white scales along the streaks.

> Similar lesions occur on the peduncles and pedicels of the panicle.

> The severity of the infection may not kill the plants, but individual leaves are partly or entirely browned.

> In some instances the entire top of the plant may be killed, tissue becomes soft and brown, especially where partly enclosed and protected by lower leaves and sheaths.

> In such case, new shoots or tillers may come out at the base.

Bacterium: *Pseudomonas avenae*

Management

> Crop rotation for 2-3 years

Chapter - 13

Diseases of Little millet / Samai - *Panicum sumatrense* Roth ex Roem. & Schult.

S. No.	Disease	Pathogen
1.	Grain smut	*Sphacelotheca destruens*
2.	Rust	*Uromyces linearis*
3.	Udbatta	*Ephelis oryzae*
4.	Downy mildew	*Sclerospora graminicola*

1. Grain smut

Symptoms

» The entire inflorescence is converted into a sorus containing spores (Chlamydospores) and fibrous vascular bundles.

» The sorus is covered by a white or grey membrane.

» Abnormal development of hairs is evident on the leaf sheaths of infected plants.

» The smut spores are round or angular and yellowish brown.

» The spores are easily blown away leaving nothing inside the glumes.

» Some of the grains developed late, remain greenish and increase in size slightly over the normal grains.

Diseases of Little millet

» Such grains release spores on pressing

Fungi: *Sphacelotheca destruens*; *Macalpinomyces sharmae* K. Vanky

Disease cycle

» Externally seed-borne and survive for more than 8 years.

Management

» Grow resistant genotypes like Varisukdhara, RLM 13, RLM 14, OLM 203, VMLC 281, VMLC 296, OLM 40, DPI 2394, PLM 202, OLM 203, DPI 2386 and CO 2.

» Seed treatment with carboxin or carbendazim at 2 g/ kg.

» Seed treatment with carboxin at 2 g/ kg followed by spray of carbendazim 0.05%.

2. Rust

Symptoms

» Numerous, minute, narrow brown pustules arranged in linear rows appear on the upper surface of the leaves.

» The brown uredia develop on upper surface of the leaves.

Fungus: *Uromyces linearis* Berk. and Br.

Management

» Seed treatment with carboxin or carbendazim at 2 g/ kg.

3. Udbatta disease

Symptoms

» The diseased plants are conspicuous by their malformed inflorescence bearing greyish white fructifications of the fungus.

» In the infected panicles, the spike lets are found to be glued to one another and to the main rachis by the viscid spore masses, which harden into a crust.

» Black sclerotial masses are formed on mature panicles.

» The inflorescence of the healthy plant is a loose panicle measuring 30 to 40 cm long where as in the diseased plant, the spike lets become glued into a cylindrical structure and the length of the panicle gets reduced to 15-23 cm long.

Fungus: *Ephelis oryzae* Syd. (Teleomorph: *Balansia oryzae-sativae* Hoshioka).

Management

» Seed treatment with carboxin 2 g/ kg.

4. Downy mildew: *Sclerospora graminicola* (Sacc.) Schroet.

» The common symptoms are downy growth on the leaves and leaf shredding, the green-ear symptoms being mostly absent.

Minor diseases

» **Leaf blight**: *Drechslera* state of *Cochliobolus nodulosus* Luttrell = *D. nodulosa* (Berk. & Curt.) Subram. & Jain (*Helminthosporium nodulosum* Berk & Curt. apud Sacc.).

» **Sheath blight:** *Rhizoctonia solani* Kuhn

Chapter - 14

Diseases of Job's tear / Kuratti paci - *Coix lacryma-jobi* L.

1. **Rust:** *Puccinia operata* Mundk. & Thirum.

   » In the leaves very minute straw colored scattered pustules are observed on both surfaces.

2. **Smut:** *Ustilago coicis* Bref. & *Ustilago lachrymae-jobi* Bref.

   » Affected ovaries of the inflorescens are transformed into black spore masses, covered by hard floral glumes. However male flowers are not attacked.

3. **Black spot:** *Phyllachora coices* Henn.

4. **Leaf blight:** *Pseudocochliobolus nisikadoi* Tsuda, Ueyama et Nishihara

Chapter - 15

Diseases of Redgram / Pigeon pea / Arhar - *Cajanus cajan* (L.) Millsp.

S. No.	Disease	Pathogen
1.	Wilt	*Fusarium udum*
2.	Dry root rot	*Macrophomina phaseolina*
3.	Powdery mildew	*Leveillula taurica*
4.	Stem blight	*Phytophthora drechsleri* f.sp. *cajani*
5.	Leaf spot	*Cercospora indica*
6.	Sterility mosaic disease	*Pigeonpea sterility mosaic virus*

1. Wilt

Symptoms

- » The disease may appear from early stages of plant growth (4-6 week old plant) up to flowering and podding.

- » The disease appears as gradual withering and drying of plants.

- » Yellowing of leaves and blackening of stem starting from collar to branches which gradually result in drooping and premature drying of leaves, stems, branches and finally death of plant.

- » Vascular tissues exhibit brown discoloration.

Diseases of Redgram

» Often only one side of the stem and root system is affected resulting in partial wilting.

Fungus: *Fusarium udum* (Butler) Snyder and Hansen.
       Earlier: *Fusarium oxysporum* f. sp. *udum*

The fungus produces hyaline, septate mycelium. Microconidia are hyaline, small, elliptical or curved, single celled or two celled. Macroconidia are also hyaline, thin walled, linear, curved or fusoid, pointed at both ends with 3-4 septa. The fungus also poduce thick walled, spherical or oval, terminal or intercalary chlamydospores singly or in chains of 2 to 3.

Disease cycle

The fungus survives in the infected stubbles in the field. The primary spread is by soil-borne chlamydospores and also by infected seed. Chlamydospores remain viable in soil for 8-20 years. The secondary spread in the field is through irrigation water and implements.

Favourable conditions

» Soil temperature of 17-25°C.

» Continuous cultivation of redgram in the same field.

Management

» Grow resistant varieties like Sharad, Jawahar, Maruthi, Malviya Arhar 2, C 11, Pusa 9, Pusa Ageti, Narendra Arhar 1, Birsa Arhar 1, AWR 74125, BDN 1, BDN 2, BWR 370, R 60, C11, NP 15, T 21, C 36 and NP (WR) 15.

» Seed treatment with *Trichoderma asperellum* at 4 g/ kg (10^6 cfu/ g).

» Avoid successive cultivation of red gram in the same field and rotation with tobacco.

» Mixed cropping with sorghum in the field.

2.   Dry root rot

Symptoms

» The disease occurs both in young seedlings and grown up plants.

» Infected seedlings can show reddish brown discoloration at collar region.

164   Diseases of Field Crops and their Management

» The lower leaves show yellowing, drooping and premature defoliation.

» The discolored area later turns to black and sudden death of the plants occurs in patches.

» The bark near the collar region shows shredding.

» The plant can be easily pulled off leaving dark rotten root in the ground.

» Minute dark sclerotia are seen in the shredded bark and root tissues.

» Large number of brown dots seen on the stem portion represents the pycnidial stage of the fungus.

Fungus: *Macrophomina phaseolina* (Tassi) Goid. (Pycnidial stage)

               Rhizoctonia bataticola (Taub.) Butler (Sclerotial stage)

The fungus produces dark, brown, filamentous hyphae and constrictions are seen in hyphal branches at the junction with main hyphae. Sclerotia are jet black, smooth, hard, minute, globose and 110-130μm in diameter. The pycnidia are dark brown and ostiolated. Conidiophores (phialides) are hyaline, short, obpyriform to cylindrical, develop from the inner walls of the pycnidium. The pycnidiospores are hyaline, single celled and ellipsoid to ovoid.

Disease cycle

The primary spread of the disease is by seed and soil. Secondary spread is by air-borne conidia. The pathogen survives as sclerotia in the soil as facultative parasite and in dead host debris.

Favourable conditions

» Prolonged drought followed by irrigation.

» High temperature of 28-35°C.

Management

» Seed treatment with carbendazim at 2g/ kg or pellet the seeds with *Trichoderma asperellum* at 4 g/ kg (10^6 cfu/ g).

» Apply heavy doses of farm yard manure or green leaf manure like *Gliricidia maculata* at 10 t/ ha or apply Neemcake at 150 kg/ ha.

Diseases of Redgram 165

3. Powdery mildew

Symptoms

> White powdery growth of the fungus can be seen on the lower surface of leaves.

> The corresponding areas in upper surface show pale yellow discoloration.

> The white powdery mass consists of conidiophores and conidia of the fungus.

> In severe cases, the white growth can be seen on the upper surface also.

> The severe infection of the fungus leads to premature shedding of leaves and plant remains barren.

Fungus: *Leveillula taurica* (Lév.) G. Arnaud, (1921)

The fungus is intercellular and absorbs nutrition through haustoria. The conidiophores, which arise through stomata, are hyaline, long, non septate, slender and rarely branched and bear single conidium at the tip. The conidia are hyaline, single celled and elliptical or clavate. The fungus also produces black, globose cleistothecia with simple myceloid appendages. They contain 9-20 cylindrical asci. Each ascus contains 3-5 ascospores which are also hyaline and unicellular.

Disease cycle

The fungus survives in the soil through cleistothecia and ascospores from asci infect the first lower most leaves near the soil level. Secondary spread is by air-borne conidia.

Favourable conditions

> Dry humid weather following rainfall.

Management

> Spray carbendazim 500 g/ ha or wettable sulphur 2 kg / ha or penconazole 10% EC 0.5 ml/ lit at the initiation of the disease and repeat after 15 days.

166 Diseases of Field Crops and their Management

4. Stem blight

Symptoms

» Initially purple to dark brown necrotic lesions girdle the basal portion of the stem and later may occur an aerial parts, lesions are small and smooth, later enlarging and slightly depressed.

» Infected tissues become soft and whole plant dies.

» In grown up plants, infection is mostly confined to basal portions of the stem.

» The infected bark becomes brown and the tissue softens causing the plant to collapse.

» In leaf, localized yellowing starts from the tip and margin and gradually extends towards the mid-rib.

» The centre of the spots later turns brown and hard.

» The spots increase in size and cover a major portion of the lamina, leading to drying.

Chromista: *Phytophthora drechsleri* f. sp. *cajani*

The pathogen produces hyaline, coenocytic mycelium. The sporangiophores are hyaline bearing ovate or pyriform, non-papillate sporangia. Each sporangium produces 8-20 zoospores. Oospores are globose, light brown, smooth and thick walled.

Disease cycle

The fungus survives in the soil and plant debris in the form of oospores. Primary infection is from oospores and secondary spread of the disease by zoospores from sporangia. Rain splash and irrigation water help for the movement of zoospores.

Favourable conditions

» Heavy rain during the months of July- September

» High temperature (28-30°C) and soils with poor drainage.

Management

» Grow resistant varieties like ICPL 38, ICPL 4135, ICPL 8564, ICPL 8610 and ICPL 8692.

Diseases of Redgram 167

» Seed treatment with metalaxyl at 6 g/ kg and spray metalaxyl at 500 g/ ha.

» Adjust the sowing time so that crop growth should not coincide with heavy rainfall.

5. Leaf spot

Symptoms

» Small, light brown coloured spots appear on leaves.

» The spots later become dark brown and the infected portions drop off leaving shot hole symptoms.

» When several spots join together, irregular necrotic blotches develop and premature defoliation occurs.

» In severe cases, black lesions develop on petioles and stem.

Fungi: *Cercospora indica* U.P. Singh; *Cercospora cajani* Henn.

The fungus produces large number of whip-like, hyaline, 7-9 septate conidia in groups on the conidiophores which are light to dark brown in colour.

Disease cycle

The fungus survives in the infected plant tissues. The disease is spread by airborne conidia.

Management

» Remove the infected plant debris and destroy.

» Spray mancozeb 2 kg or carbendazim 500 g/ ha soon after the appearance of symptom and repeat after a fortnight.

6. Sterility mosaic disease (SMD)

Symptoms

» The symptoms are characterized by bushy and pale green appearance of plants.

» The excessive vegetative growth, stunting, prominent mosaic on leaves and reduction in leaf size.

168   Diseases of Field Crops and their Management

» Complete or partial cessation of flowering leads to sterility.

» Depending on genotype three types of symptoms are recognized. They are

     a. Severe mosaic and sterility
     b. Mild mosaic and partial sterility
     c. Chlorotic ringspot without any noticeable sterility.

Virus: *Pigeonpea sterility mosaic virus* (PPSMV).

Disease cycle

It is not transmitted by infectious sap. It is transmitted by an eriophyid mite, *Aceria cajani* in a semi persistant manner, mites retaining the virus 12-13 hours, eggs of mites do not transmit. The self grown redgram plants and perennial species act as source of virus inoculums.

Management

» Rogue out infected plants up to 40 days after sowing.

» Grow resistant genotypes/varieties like ICP 7035, VR3, Purple 1, DA11, DA32, ICP 6997, Bahar, BSMR 235, ICP 7198, PR 5149, ICP 8861, Pusa 15, Pusa 17, Pusa 18 and BSR 1.

» Spray monocrotophos at 500 ml/ ha soon after appearance of the disease and if necessary, repeat after 15 days.

There are two other virus diseases reported on pigeonpea, mosaic and yellow mosaic transmitted by aphids and whiteflies which are of sporadic occurrence only.

Minor diseases

Seedling blight: *Sclerotium rolfsii* Sacc.

» Small brown water soaked dots appear near collar region, expands to irregular necrotic spots leading to girdling of stem and death of seedling.

Brown blotch: *Colletotrichum capsici* (Syd.) E.J. Butler & Bisby, (1931)

» Purple brown discolouration occurs mainly on pods but also on petioles, leaf veins, stems and peduncles. Pods become distorted and have black fruiting bodies.

Diseases of Redgram

Anthracnose: Teleomorph: *Glomerella cingulata* (Stoneman) Spauld. & H. Schrenk)

Anamorph: *Colletotrichum lindemuthianum* (Sacc. & Magnus) Briosi & Cavara, (1889);

» Black lesions develop on stem which spreads to leaf petiole and leaves. Black sunken lesions also develop on pod.

Stem rot: *Pythium aphanidermatum* (Edson) Fitzp., (1923)

» Seedlings of 2-3 weeks old are severely attacked at collar region and death occurs immediately. Greyish green water soaked lesions develop on adult plants, leading to girdling of stem.

Leaf spot: *Alternaria alternata* (Fr.) Keissl. (1912)

» Water soaked, circular to irregular spots occur. The centre of the spot is straw coloured with raised reddish brown margins.

Halo blight: *Pseudomonas phaseolicola*

» Small brown spots appear on leaves and develop a chlorotic halo.

» The spots extend and form dried brown zone. Brown elongated streaks appear on petioles, stem and pods.

Chapter - 16

Diseases of Black gram / Urd bean - *Vigna mungo* (L.) Hepper and Diseases of Green gram / Mungbean - *Vigna radiata* (L.) R. Wilczek

S. No.	Disease	Pathogen
1.	Powdery mildew	*Erysiphe polygoni*
2.	Anthracnose	*Colletotrichum lindemuthianum*
3.	Leaf spot	*Cercospora canescens*
4.	Rust	*Uromyces phaseoli* var. *typica*
5.	Root rot and leaf blight	*Rhizoctonia bataticola*
6.	Mungbean yellow mosaic disease	*Mungbean yellow mosaic virus*
7.	Leaf crinkle disease	*Urdbean leaf crinkle virus*
8.	Leaf curl	*Groundnut bud necrosis virus*

1. Powdery mildew

Symptoms

» Small, irregular powdery spots appear on the upper surface of the leaves, sometimes on both the surfaces.

» The disease becomes severe during flowering and pod development stage.

» The white powdery spots completely cover the leaves, petioles, stem and even the pods.

Diseases of Black gram and Green gram

» The plant assumes greyish white appearance; leaves turn yellow and finally shed.

» Often pods are malformed and small with few ill-filled seeds.

Fungus: *Erysiphe polygoni* DC.

The fungus is ectophytic, spreading on the surface of the leaf, sending haustoria into the epidermal cells. Conidiophores arise vertically from the leaf surface, bearing conidia in short chains. Conidia are hyaline, thin walled, elliptical or barrel shaped or cylindrical and single celled. Later in the season, cleistothecia appear as minute, black, globose structures with myceloid appendages. Each cleistothecium contains 4-8 asci and each ascus contains 3-8 ascospores which are elliptical, hyaline and single celled.

Disease cycle

» The Pathogen is an obligate parasite and survives as cleistothecia in the infected plant debris. Primary infection is usually from ascospores from perennating cleistothecia. The secondary spread is carried out by the air-borne conidia. Rain splash also helps in the spread of the disease.

Favourable conditions

» Warm humid weather.

» The disease is severe generally during late kharif and rabi seasons.

Management

» Remove and destroy infected plant debris.

» Spray carbendazim 500 g or wettable sulphur 2 kg/ ha at the initiation of disease and repeat 15 days later.

» Spray with penconazole 10% EC at 0.5 ml/ lit.

2. Anthracnose

Symptoms

» The symptom can be observed in all aerial parts of the plants and at any stage of crop growth.

» The fungus produces dark brown to black sunken lesions on the hypocotyl area and cause death of the seedlings.

» Small angular brown lesions appear on leaves, mostly adjacent to veins, which later become greyish white centre with dark brown or reddish margin.

» The lesions may be seen on the petioles and stem. The prominent symptom is seen on the pods.

» Minute water soaked lesion appears on the pods initially and becomes brown and enlarges to form circular, depressed spot with dark centre with bright red or yellow margin.

» Several spots join to cause necrotic areas with acervuli. The infected pods have discolored seeds.

Fungus: Anamorph: *Colletotrichum lindemuthianum* (Sacc. & Magnus) Briosi & Cavara; Teleomorph: *Glomerella lindemuthianum* Shear (1913)

The fungus mycelium is septate, hyaline and branched. Conidia are produced in acervuli, arise from the stroma beneath the epidermis and later rupture to become erumpent. A few dark coloured, septate setae are seen in the acervulus. The conidiophores are hyaline and short and bear oblong or cylindrical, hyaline, thinwalled, single celled conidia with oil globules. The perfect stage of the fungus produces perithecia with limited number of asci, which contain typically 8 ascospores which are one or two celled with a central oil globule.

Disease cycle

The fungus is seed-borne and cause primary infection. It also lives in the infected plant tissues in soil. The secondary spread by air borne conidia produced on infected plant parts. Rain splash also helps in dissemination.

Favourable conditions

» High relative humidity (> 90%) and low temperature 15-20°C

» Cool rainy days.

Management

» Remove and destroy infected plant debris in soil.

Diseases of Black gram and Green gram

» Seed treatment with carbendazim at 2 g/ kg.

» Spray carbendazim 500 g or mancozeb 2 kg/ ha soon after the appearance of disease and repeat after 15 days.

3. Leaf spot

Symptoms

» Small, circular spots develop on the leaves with grey centre and brown margin.

» Several spots coalesce to form brown irregular lesions.

» In severe cases defoliation occurs.

» The brown lesions may be seen on petioles and stem in severe cases.

» Powdery growth of the fungus may be seen on the centre of the spots.

Fungus: *Cercospora canescens* Ellis & G. Martin

The fungus produces clusters of dark brown septate conidiophores. The conidia are linear, hyaline, thin walled and 5-6 septate.

Disease cycle

The fungus survives on diseased plant debris and on seeds. The secondary spread is by air-borne conidia.

Favourable conditions

» Humid weather and dense plant population.

Management

» Remove and burn infected plant debris.

» Varieties of Green gram AAU 34 and AAU 39 are tolerant to leaf spot.

» Spray Mancozeb at 2 kg/ ha or Carbendazim at 1g/ l or Chlorothalanil 1.5g/ l.

4. Rust

Symptoms

» The disease is mostly seen on leaves, rarely on petioles, stem and pods.

» The fungus produces small, round, reddish brown uredosori mostly on lower surface.

» They may appear in groups and several sori coalesce to cover a large area of the lamina.

» In the late season, teliosori appear on the leaves which are linear and dark brown in colour.

» Intense pustule formation causes drying and shedding of leaves.

Fungus: *Uromyces phaseoli* var. *typica* Arthur
Syn: *Uromyces appendiculatus* (Pers.) Link 1816

It is autoecious, long cycle rust and all the spore stages occur on the same host. The uredospores are unicellular, globose or ellipsoid, yellowish brown with echinulations. The teliospores are globose or elliptical, unicellular, pedicellate, chestnut brown in colour with warty papillae at the top. Yellow coloured pycnia appear on the upper surface of leaves. Orange coloured cupulate aecia develop later on the lower surface of leaves. The aeciospores are unicellular and elliptical.

Disease cycle

The pathogen survives in the soil through teliospores and as uredospores in crop debris. Primary infection is by the sporidia developed from teliospores. Secondary spread is by wind-borne uredospores. The fungus also survives on other legume hosts.

Favourable conditions

» Cloudy humid weather, temperature of 21-26°C.

» Nights with heavy dews.

Management

» Remove the infected plant debris and destroy.

Diseases of Black gram and Green gram

» Spray mancozeb 2 kg or carbendazim 500 g or propiconazole 1lit/ ha, immediately on the set of disease and repeat after 15 days.

5. Root rot and leaf blight

Symptoms

a. Seedling blight

» The leaves of green gram are blightened and dry-off.

» Infected leaves are binded one over another by the mycelium.

b. Leaf blight

» The disease symptom starts initially with yellowing and drooping of the leaves.

» The leaves later fall off and the plant dies with in week.

c. Web blight or aerial blight

» Infection begins in the lower most leaves of green gram near to soil.

» Reddish brown pappery lesions are seen on leaves.

» Infected leaves are binded one over another by the mycelium.

» The leaves will fall off.

» On young pods the spots are light tan and irregular in shape but generally on mature pods, they are dark brown and sunken.

d. Dry root rot

» Dark brown lesions are seen on the stem at ground level and bark shows shredding symptom.

» The affected plants can be easily pulled out leaving dried, rotten root portions in the ground.

» The rotten tissues of stem and root contain a large number of black minute sclerotia.

Fungus: *Rhizoctonia bataticola* (Taub.) Butler; *Rhizoctonia solani* Kuhn (Green gram) *Macrophomina phaseolina* (Tassi) Goid. (Pycnidial stage)

The fungus produces dark brown, septate mycelium with constrictions at hyphal branches. Minute, dark colored, round sclerotia in abundance. The fungus also produces dark brown, globose ostiolated pycnidia on the host tissues. The pycnidiospores are thin walled, hyaline, single celled and elliptical.

Disease cycle

The fungus survives in the infected debris and also as saprophyte in soil. The primary spread is through seed-borne and soil-borne sclerotia. The secondary spreads is through pycnidiospores which are air-borne.

Favourable conditions

- » Day temperature of 30°C.
- » Prolonged dry season followed by irrigation.

Management

- » Apply farm yard manure or green leaf manure (*Gliricidia maculata*) at 10 t/ha or neemcake at 150 kg/ ha.
- » Seed treatment with carbendazim + azoxystrobin at 2 g/ kg (1:1 ratio) or pellet the seeds with *Trichoderma asperellum* at 4 g/ kg (10^6 cfu/ g) or *Pseudomonas fluorescens* at (10^6 cfu/ g).

6. Mungbean yellow mosaic disease

Symptoms

- » Initially small yellow patches or spots appear on green lamina of young leaves.
- » Soon it develops into a characteristics bright yellow mosaic or golden yellow mosaic symptom.
- » Yellow discoloration slowly increases and leaves turn completely yellow.
- » Infected plants mature later and bear few flowers and pods.
- » The pods are small and distorted. Early infection causes death of the plant before seed set.

Virus: *Mungbean yellow mosaic virus* (MYMV)

Diseases of Black gram and Green gram

It is caused by *Mungbean yellow mosaic India virus* (MYMIV) in Northen and Central region and *Mungbean yellow mosaic virus* (MYMV) in western and southern regions. It is a Begomovirus belonging to the family geminiviridae. Geminate virus particles, ssDNA, bipartite genome with two gemonic components DNA-A and DNA-B.

Disease cycle

Transmitted by whitefly (*Bemisia tabaci*). Disease spreads by feeding of plants by viruliferous whiteflies. Summer sown crops are highly susceptible. Weed hosts *viz.*, *Croton sparsiflorus, Acalypha indica, Eclipta alba* and other legume hosts serve as reservoir for inoculum.

Management

» Rogue out the diseased plants up to 40 days after sowing.

» Remove the weed hosts periodically.

» Increase the seed rate 25 kg/ ha.

» Grow resistant black gram variety like VBN 1, PDU 10, IC 12/2 and PLU 322.

» Follow mixed cropping by growing two rows of maize (60 × 30 cm) or sorghum (45 × 15 cm) or cumbu (45 × 15 cm) for every 15 rows of black gram or green gram.

» Seed treatment with thiomethoxam 70 WS or imidacloprid 70 WS at 4 g/ kg.

» Spray *Paecilomyces farinosus* for whitefly control.

» Spray thiamethoxam 25 WG at 100 g or imidacloprid 17.8% SL at 1 ml/ 5 lit.

7.   Leaf crinkle disease

Symptoms

» Crinkling and curling of the tips of leaflets and increase in leaf area.

» Crinkling and rugosity in older leaves becomes severe and leaves thickened.

» Petioles as well as internodes are shortened.

» Infected plant gives a stunted and bushy appearance.

» Flowering is delayed, if inflorescence is formed, is malformed with small size flower buds and fails to open.

Virus: *Urdbean leaf crinkle virus* (ULCV)

Disease cycle

Presence of weed hosts like *Aristolochia bracteata* and *Digera arvensis.* Kharif season crop and continuous cropping of other legumes serve as source of inoculum. The virus is seed-borne and primary infection occurs through infected seeds. Perhaps white fly, *Bemisia tabaci* helps in the secondary spread. The virus is also sap transmissible.

Management

- » Use increased seed rate 25 kg/ ha.
- » Rogue out the diseased plants at weekly interval up to 45 days after sowing.
- » Remove weed hosts periodically.
- » Spray methyl demeton at 500 ml/ ha on 30 and 40 days after sowing.

8. Leaf curl / Necrosis

Symptoms

- » Upward cupping and curling of leaves with vein clearing.
- » Infected leaves turn brittle and sometimes show vein necrosis on the under surface of the leaves, extending to the petiole.
- » Plants affected in the early stages of growth develop top necrosis and die.
- » Plant may produce a few small and malformed pods.

Virus: *Groundnut bud necrosis virus*

Disease cycle

The virus is transmitted by thrips *viz., Frankliniella schultzii, Thrips tabaci* and *Scirtothrips dorsalis.* The virus survives in weed hosts, tomato, petunia and chilli.

Management

- » Rogue out infected plants up to 30 days after sowing.

Diseases of Black gram and Green gram

» Remove the weed hosts which harbour virus and thrips.

» Spray imidachlor at 500 ml/ ha on 30 and 45 days after sowing.

Minor diseases

***Ascochyta* leaf spot:** *Ascochyta phaseolorum* Sacc. 1878.

» Small irregular spot with grey to brown centre and yellow border.

» They rapidly enlarge to produce very large brown lesions with concentric markings.

Bacterial blight: *Xanthomonas phaseoli* (ex Smith 1897) Gabriel at al. 1989

» Circular, reddish brown spots appear on leaves, enlarge to form irregular brown lesions.

» Water soaked, sunken spots with red border occur on pods.

Chapter - 17

Diseases of Bengal gram / Chickpea - *Cicer arietinum* L.

S. No.	Disease	Pathogen
1.	*Ascochyta* blight	*Ascochyta rabiei*
2.	Rust	*Uromyces ciceris-arietini*
3.	Wilt	*Fusarium oxysporum* f. sp. *ciceris*
4.	Collar rot	*Sclerotium rolfsii*
5.	Stunt disease	*Chick pea chlorotic dwarf virus*

1. *Ascochyta* blight

Symptoms

» All above ground parts of the plant are infected.

» On leaf, the lesions are round or elongated, bearing irregularly depressed brown spot and surrounded by a brownish red margin.

» Similar spots may appear on the stem and pods.

» The spots on the stem and pods have pycnidia arranged in concentric circles as minute block dots.

» When the lesions girdle the stem, the portion above the point of attack rapidly dies.

Diseases of Bengal gram

» If the main stem is girdles at the collar region, the whole plant dies.

Fungus: *Ascochyta rabiei* (Pass.) Labrousse

The fungus produces hyaline to brown and septate mycelium. Pycnidia are spherical to sub-globose with a prominent ostiole. Pycnidiospores are hyaline, oval to oblong, straight or slightly curved and single celled, occasionally bicelled.

Disease cycle

The fungus survives in the infected plant debris as pycnidia. The pathogen is also externally and internally seed-borne. The primary spread is from seed-borne pycnidia and plant debris in the soil. The secondary spreads is mainly through air-borne pycnidiopores (conidia). Rain splash also helps in the spread of the disease.

Favourable conditions

» High rainfall during flowering.

» Temperature of 20-25°C and relative humidity of 60%.

Management

» Grow resistant varieties like C 235, Gwalior 3, B.G.-408, Vishwas, GNG 146, H 75, GG 588, ILC 72, ILC 192, ILC 201, ILC 202, ICC 4200, ICC 4248 and ICC 5124.

» Remove and destroy the infected plant debris in the field.

» Follow crop rotation with cereals.

» Seed treatment with carbendazim + azoxystrobin (1:1 ratio) at 2 g/ kg.

» Seed treatment with *Pseudomonas fluorescens* 4 g/ kg.

» Exposure of seed at 40-50°C reduced the survival of *A. rabiei* by about 40-70%.

» Spray with carbendazim at 500 g/ ha or chlorothalonil 1 kg/ ha.

2. Rust

Symptoms

» The infection appears as small oval, brown, powdery lesions on both the surface, especially more on lower surface or leaf.

» The lesions, which are uredosori, cover the entire leaf surface.

» Late in the season dark teliosori appear on the leaves.

» The rust pustules may appear on petioles, stems and pods.

Fungus: *Uromyces ciceris-arietini* Jacz.

The uredospores are spherical, brownish yellow in colour, loosey echinulated with 4-8 germ pores. Teliospores are round to oval, brown, single celled with unthickened apex and the walls are rough, brown and warty.

Disease cycle

The fungus survives as uredospores in the legume weed *Trigonella polycerata* during summer months and serve as primary source of infection. The spread is through wind-borne uredospores.

Management

» Destory weed host.

» Spray carbendazim 1 g/ lit or propiconazole 1 g/ lit or mancozeb + metalaxyl 2 g/ lit.

3. Wilt

Symptoms

» The disease occurs at two stages of crop growth, seedling stage and flowering stage stage.

» The main symptoms on seedlings are yellowing and drying of leaves, drooping of petioles and rachis, withering of plants.

» In the case of adult plants drooping of leaves is observed initially in upper part of plant, and soon observed in entire plant.

» Vascular browning is conspicuously seen on the stem and root portion.

Fungus: *Fusarium oxysporum* f. sp. *ciceris* Matuo & K. Sato [as ciceri], (1962)

Diseases of Bengal gram

The fungus produces hyaline to light brown, septate and profusely branched hyphae. Microconidia are oval to cylindrical, hyaline, single celled, normally arise on short conidiophores. Macroconidia which borne on branched conidiophores, are thin walled, 3 to 5septate, fusoid and pointed at both ends. Chlamydospores are roughwalled or smooth, terminal or intercalary, may be formed singly or in chains.

Disease cycle

The disease is seed and soil borne. The primary infection is through chlamydospores in soil, which remain viable upto next crop season. The secondary spread is through irrigation water, cultural operations and implements.

Favourable conditions

» High soil temperature (above 25°C).

» High soil moisture.

Management

» Grow resistant cultures like C 235, G 124, B 115, BG 408, WBG 39/2, ICCC 42, H82-2, PPK 1, PPK 2, P 621, Avrodhi, Alok Samrat, Pusa 209, Pusa 212, Pusa 372, JG 322, GPF 2, Haryanachana 1 and Kabuli chickpea like Pusa 1073 and Pusa 2024.

» Apply heavy doses of organic manure or green manure.

» Seed treatment with carbendazim at 2 g/ kg with *Trichoderma asperellum* at 4 g/ kg (10^6 cfu/ g) *Pseudomonas fluorescens* at 10 g/ kg (10^6 cfu/ g).

4.   Collar rot

Symptoms

» It comes in the early stages i.e. up to six weeks from sowing.

» Drying plants whose foliage turns slightly yellow before death, scattered in the field is an indication of the disease.

» Seedlings become chlorotic.

» The joint of stem and root turns soft slightly contracts and begins to decay.

» Infected parts turn brown white.

> Black dots, like mustard in shape known as sclerotia are seen appearing on the white infected plant parts.

Fungus: *Sclerotium rolfsii* Sacc.

Favorable conditions

> High soil moisture, low soil pH and high temperature.

> The presence of undecomposed organic matter on the soil surface and high moisture at the time of sowing and at the seedling stage

> Disease incidence is higher when sown after rice or early sown crop.

Management

> Deep ploughing in summer.

> Avoid high moisture at the sowing time.

> Seedlings should be protected from excessive moisture.

> Destroy the crop residues of last crop and weeds before sowing and after harvest.

> All undecomposed matter should be removed from the field before land preparation.

> Seed treatment with a mixture of carbendazim + azoxystrobin (1:1) at 2 g/ kg.

5.   Stunt disease

Symptoms

> Affected plants are stunted and bushy with short internodes.

> The leaflets are smaller with yellow, orange or brown discoloration.

> Stem also shows brown discoloration.

> The plants dry prematurely.

> If survive, a very few small pods are formed.

> Phloem browning in the collar region is the most characteristic symptom of the stunt, leaving xylem normal.

Diseases of Bengal gram 185

Viruses: *Chick pea chlorotic dwarf virus* (CpCDV) and *Maize streak virus* (MSV).

Disease cycle

» The virus is transmitted by *Aphis craccivora.*

Management

» Rogue out the infected plants.

» Spray monocrotophos at 500 ml/ ha.

Minor diseases

Foot rot: *Operculella padwickii* Khesw., (1941)
Syn: *Phacidiopycnis padwickii* (Khesw.) B. Sutton, (1980)

» Rotting is evident from collar region onwards.

» Internal brown discolouration appears above the rotton portion (only on bark portion).

Stemrot: *Sclerotinia sclerotiorum* (Lib.) de Bary (1884)

» The disease appears mostly on stems rot of adult plants as water soaked lesion on upper parts of stem.

» The affected portion is covered with white cottony growth and black sclerotial bodies.

Bacterial leaf blight: *Xanthomonas campestris* pv. *cassiae*

» Small water soaked lesions develop on leaves with chlorotic haloes which later turn to dark brown spots.

» Post emergence seedling rot is also common.

Bean common mosaic virus

» Infect plants are stunted, bushy appearance of plant with mosaic mottling.
Vector: *Aphis gossypii* and *A. craccivora.*

Chapter - 18

Diseases of Horsegram - *Macrotyloma uniflorum* (Lam.) Verdc.

S. No.	Disease	Pathogen
1.	Anthracnose	*Colletotrichum lindemuthianum*
2.	Root rot	*Pellicularia koleroga*
3.	Dry root rot	*Macrophomina phaseolina*
4.	Aerial blight	*Rhizoctonia solani*
5.	Powdery mildew	*Leveillula taurica*
6.	Rust	*Uromyces phaseoli-typica*
7.	*Cercospora* leaf spot	*Cercospora dolichos*
8.	Bacterial leaf spot	*Xanthomonas phaseoli* var. *sojansis*
9.	Mosaic complex	*Horsegram yellow mosaic virus*

1. Anthracnose

Symptoms

- » Symptoms are circular, black, sunken spots with dark center and bright red orange margins on leaves and pods.

- » In severe infections, the affected parts wither off.

Diseases of Horsegram

» Seedlings get blighted due to infection soon after seed germination

Fungus: *Colletrotrichum lindemuthianum* (Sacc. Angn.) Bri. and Cov.

Disease cycle

» The pathogens survive on seed and plant debris

» Disease spreads in the field through air-borne conidia

Favourable conditions

» The disease is more severe in cool and wet seasons.

Management

» Early planting i.e. immediately after onset of monsoon.

» Grow crop on bower system to avoid soil contact.

» Maintain proper drainage in the field.

» Spray chlorothalonil 0.2%.

2. Root rot

Symptoms

» The pathogens cause seed decay, root rot, damping-off, seedling blight, stem canker and leaf blight in horsegram

» The disease occurs commonly at pod development stage

» The affected leaves turn yellow in colour and brown irregular lesions appear on leaves.

» Roots and basal portion of the stem become black in colour and the bark peels off easily.

» When the tap root of the affected plant is split open, reddening of internal tissues is visible.

Fungi: *Pellicularia koleroga* Cooke; *Pellicularia filamentosa* (Pat.) Rogers

Disease cycle

» Species are saprotrophic, occurring in the soil which is the source of primary infection.

» Secondary infection occurs by means of asexual spores.

Favourable conditions

» Moist soil and humid conditions favour the development of disease.

Management

» Plant in well-draining soils.

» Prepare seed beds to enhance rapid germination

» Proper irrigation is provided to prevent flooding and saturated soil conditions.

» Soil amendment with farm yard manure at 5 tonnes/ acre

3. Dry root rot

Symptoms

» The first outward symptom of the disease is yellowing of the leaves.

» Within three or four days they drop off.

» The affected plants wilt and die within a week.

» The bark of the root and basal stem are fibrous and are found associated with black powdery mass of sclerotia of the fungus.

» The plants bear pods with partially filled seeds. The disease appears in patches and becomes severe during dry periods.

Fungus: *Macrophomina phaseolina* (Tassi) Goud.

Favourable conditions

» Dry sandy soil and low humid conditions favour the development of disease.

» Management

Diseases of Horsegram

» Application of organic amendments like farmyard manure and neem cake.

» Enhanced soil moisture can reduce the *M. phaseolina* population.

» Seed treatment with *Trichoderma viride* 4 g/ kg and *Pseudomonas fluorescens* 10 g/ kg.

4. Aerial blight

Symptoms

» Aerial blight is seen on the foliage as irregular water soaked area.

» Under high atmospheric humidity the spots coalesce rapidly and cover a large part of the leaf lamina.

» There will be white mycelial growth also on the affected area.

» Severely affected leaves shed in large number.

Fungus: *Rhizoctonia solani* Kuhn.

Favourable conditions

» Dry sandy soil and low humid conditions favour the development of disease.

Management

» Soil application of *Trichoderma viride* or *T. harzianum*

5. Powdery mildew

Symptoms

» White powdery patches appear on leaves and other green parts which later become dull coloured

» These patches gradually increase in size and become circular covering the lower surface

» In severe infection, both the surfaces of the leaves are completely covered by whitish powdery growth.

» In severe infections, foliage becomes yellow causing premature defoliation.

» The disease also creates forced maturity of the infected plants which results in heavy yield losses.

Fungus: *Leveillula taurica* (Lev.) Arnaudi

Disease cycle

» The pathogen has a wide host range and survives in oidial form on various hosts in offseason

» Secondary spread is through air-borne oidia produced in the season.

Favourable conditions

» Dry and moist weather, relative humidity of 90% favours disease development.

Management

» Provide proper spacing

» Spray wettable sulphur 0.2% or calixin 0.02%.

6. Rust

Symptoms

» The fungus produces characteristic rust pustules on the plant.

» The pustules are mostly found on the leaf blade.

» They are more conspicuous on the under surface of the leaves.

» Often a number of such pustules may occur on the same leaf.

» The infection may spread to young stem also.

» Plants with heavy rust infection will give a brown tinge when looked from a distance.

» In advanced stage of infection the leaf may wither resulting in considerable damage to the crop.

Fungus: *Uromyces phaseoli-typica* Arth.
Syn: *Uromyces appendiculatus* (Pers) Fries.

Diseases of Horsegram 191

Management

- » Provide proper spacing
- » Spray triadimenol 0.2%.

7. *Cercospora* leaf spot

Symptoms

- » Moist weather and splashing rains are conducive to disease development. Most outbreaks of the disease can be traced back to heavy rainstorms that occur in the area.
- » Infected leaves show small, brown, water soaked, circular spots surrounded with yellowish halo.
- » On older plants, the leaflet infection is mostly on older leaves and may cause serious defoliation. The most striking symptoms are on the green fruit.
- » Small, water-soaked spotsfirst appear which later become raised and enlarge until they are one-eighth to one-fourth inch in diameter.
- » Centres of these lesions become irregular, light brown and slightly sunken with a rough, scabby surface.

Fungus: *Cercospora dolichos* Ell. & Evir.

Disease cycle

- » The fungus survives on diseased plant debris. Fungus spreads about 3 m through the soil in one season.

Favourable conditions

- » Moist weather and splashing rains.
- » High humidity or persistent dew

Management

- » Provide proper spacing
- » Spray mancozeb 0.2%

8. Bacterial leaf spot

Symptoms

» This is a common disease of horsegram occurring on the foliage at any stage of the growth.

» The pathogen attacks the foliage causing characteristic leaf spots and blight. Early blight is first observed on the plants as small, black lesions mostly on the older foliage.

» Spots enlarge, and by the time they are one-fourth inch in diameter or larger, concentric rings in a bull's eye pattern can be seen in the center of the diseased area.

» Tissue surrounding the spots may turn yellow. If high temperature and humidity occur at this time, much of the foliage is killed.

» Lesions on the stems are similar to those on leaves, sometimes girdling the plant if they occur near the soil line.

Bacterium: *Xanthomonas phaseoli* var. *sojansis*

Disease cycle

» The bacterium is seed-borne

Favourable conditions

» Rain splashes play an important role in the development and spreadof the disease.

» Warm, rainy and wet weather is congenial.

Management

» Early planting i.e. immediately after onset of monsoon.

» Grow crop on bower system to avoid soil contact.

» Maintain proper drainage in the field.

» Spray streptomycin sulphate 300 ppm.

9. Mosaic complex

Symptoms

- » Initially mild scattered yellow spots appear on young leaves.
- » The next trifoliate leaves emerging from the growing apexshow irregular yellow and green patches alternating with each other.
- » Spots gradually increase in size and ultimately some leaves turn completely yellow.
- » Infected leaves also show necrotic symptoms.
- » Diseased plants are stunted, mature late and produce very few flowers and pods.
- » Pods of infected plants are reduced in size and turn yellow in colour.

Virus: *Horsegram yellow mosaic virus* (*Begomovirus*) (HgYMV)
Mung bean yellow mosaic virus (*Geminivirus*)

Transmission

- » The disease is transmitted in semi persistent manner by aphids and whiteflies.
- » Aphids are more active in warm summer conditions and increased their population as well as spread the viruses

Management

- » Spray neem seed kernel extract (NSKE) 5%.
- » Spray neem oil 1%

Chapter - 19

Diseases of Soybean
- *Glycine max* (L.) Merr.

S. No.	Disease	Pathogen
1.	Wilt	*Fusarium oxysporum* f. sp. *tracheiphilum*
2.	Dry root rot	*Macrophomina phaseolina*
3.	Leaf spot	*Cercospora sojana*
4.	Powdery mildew	*Microsphaera diffusa*
5.	Bacterial blight	*Pseudomonas syringae* pv. *glycinea*
6.	Bacterial pustule	*Xanthomonas axonopodis* pv. *glycines*
7.	Soybean mosaic disease	Soybean mosaic virus

1.   Wilt

Symptoms

- » Symptoms do not appear until the plants are about six weeks old.

- » Initially a few plants are noticed with pale green flaccid leaves which soon turn yellow.

- » Growth is stunted, chlorosis, drooping, premature shedding or withering of leaves with veinal necrosis often occurs and finally plant dies within 5 days.

Diseases of Soybean

> Brownish, purple discoloration of the cortical area is seen, often extends throughout the plant.

Fungus: *Fusarium oxysporum* f. sp. *tracheiphilum* (E.F. Sm.) W.C. Snyder & H.N. Hansen

The fungus produces falcate shaped macroconidia which are 4-5 septate, thin walled and hyaline. The microconidia are single celled hyaline and oblong or oval. The chlamydospores are also produced in abundance.

Disease cycle

The fungus survives in the infected stubbles in the field. The primary spread is through soilborne chlamydospores and infected seeds. The secondary spread is through conidia by irrigation water.

Favourable conditions

> Temperature of 20-25°C and moist humid weather.

Management

> Seed treatment with carbendazim at 2 g/ kg or *Trichoderma asperellum* at 4 g/ kg.

> Spot drenching with carbendazim at 0.5 g/ litre.

2. Dry root rot

Symptoms

> The disease symptom starts initially with yellowing and drooping of the leaves.

> The leaves later fall off and the plant dies with in week.

> Dark brown lesions are seen on the stem at ground level and bark shows shredding symptom.

> The affected plants can be easily pulled out leaving dried, rotten root portions in the ground.

> The rotten tissues of stem and root contain a large number of black minute sclerotia.

Fungus: *Macrophomina phaseolina* (Goid) Tassi

The fungus produces dark brown, septate mycelium with constrictions at hyphal branches. Minute, dark and round sclerotia are abundance. The fungus also produces dark brown, globose ostiolated pycnidia on the host tissues. The pycnidiospores are thin walled, hyaline, single celled and elliptical.

Disease cycle

The fungus survives in the infected debris and also as facultative parasite in soil. The primary spread is through seed-borne and soil-borne sclerotia. The secondary spread is through seed-borne and soil-borne sclerotia. The secondary spreads is through pycnidiospores which are air-borne.

Favourable conditions

» Day temperature of 30°C

» Prolonged dry season followed by irrigation.

Management

» Seed treatment with carbendazim or azoxystrobin at 2 g/ kg or pellet the seeds with *Trichoderma asperellum* at 4 g/ kg or *Pseudonomas fluorescens* at 10 g/ kg.

» Apply farm yard manure or green leaf manure (*Gliricidia maculata*) at 10 t/ ha or neem cake at 150 kg/ ha.

3. Leaf spot

Symptoms

» Light to dark gray or brown areas varying from specks to large blotches appear on seeds.

» The disease primarily affects foliage, but, stems, pods and seeds may also be infected.

» Leaf lesions are circular or angular, at first brown then light brown to ash grey with dark margins.

» The leaf spot may coalesce to form larger spots.

» When lesions are numerous the leaves wither and drop prematurely.

» Lesions on pods are circular to elongate, light sunken and reddish brown.

Diseases of Soybean

Fungus: *Cercospora sojana* K. Hara

Favourable conditions

- » Fungus survives in infected seeds and in debris.
- » Warm, humid weather favor disease incidence.

Management

- » Use resistant varieties.
- » Rotate soybean with cereals.
- » Completely remove plant residue by clean ploughing the field soon after harvest.
- » Seed treatment with azoxystrobin + carbendazium (1:1) at 2 g/ kg seed.
- » Spray mancozeb at 2g/ lit or carbendazim 0.5 g/ lit.

4. Powdery mildew

Symptoms

- » All aerial plant parts can be infected by powdery mildew, although symptoms are most easily seen on the upper surfaces of leaves.
- » Infected leaves have white to light grey, powdery patches.
- » These patches may enlarge and cover the surfaces of many leaves throughout a plant; however, infected leaves tend to be most common on the lower and middle leaves.

Fungus: *Microsphaera diffusa* Cooke & Peck

Favourable conditions

- » Cool air temperatures and low relative humidity

Management

- » Use resistant varieties.
- » Spray wettable sulphur at 2 g/ lit or thiophanate methyl 1 g/ lit.

5. Bacterial blight

Symptoms

- » Plants infected early in the growing season are characterized by brown spots on the margins of the cotyledons.
- » Young plants may be stunted, and if the infection reaches the growing point, they may die.
- » Symptoms in later growth stages include angular lesions, which begin as small yellow to brown spots on the leaves.
- » The centers of the spots will turn a dark reddish-brown to black and dry out.
- » A yellowish-green "halo" will appear around the edge of the water-soaked tissue.

Bacterium: *Pseudomonas syringae* pv. *glycinea*

Favourable conditions

- » Bacterial cells are carried by splashing or wind-driven water droplets.
- » Seedlings may be infected through infested seed.
- » Rainfall along with heavy wind.
- » Cool and wet weather.

Management

- » Use disease free seed.
- » Follow crop rotation.
- » Grow resistant varieties like Alankar, Bragg and Durga.
- » Seed treatment with cerason or captan 3 g/ kg or *Trichoderma viride* 5 g/ kg.
- » Spray copper oxychloride 0.25% or streptocyclin 500ppm.

Diseases of Soybean

6. Bacterial pustule

Symptoms

» Initial symptoms are the appearance of tiny pale green spots on leaves.

» These spots have raised centers that may develop on either surface of the leaf but are more common on the lower surface.

» Lesions are often associated with main leaf veins small, light-colored pustules will form in the center of the spots.

» Spots may merge together to form irregular lesions.

» This disease can easily be confused with soybean rust.

Bacterium: *Xanthomonas axonopodis* (syn. *campestris*) pv. *glycines*

Favourable conditions

» Wind-driven rain or water droplets.

» Warm weather with frequent showers.

Management

» Use disease free seed.

» Follow crop rotation.

» Seed treatment with cerason or captan 3g/ kg. or *Trichoderma viride* 5g/ kg.

» Spray copper oxychloride 0.25% or streptocyclin 500ppm.

7. Soybean mosaic disease

Symptoms

» Diseased plants are usually stunted with distorted (puckered, crinkled, ruffled, narrow) leaves.

» Pods become fewer and smaller seeds.

» Infected seeds get mottled and deformed.

» Infected seeds fail to germinate or they produce diseased seedlings.

Virus: *Soybean mosaic virus* (SMV)

It is caused by *Soybean mosaic virus* - a potyvirus. Flexuous particles with 750-900 nm long, ss RNA genome.

Disease cycle

Soybean mosaic virus is seed borne. The SMV can be transmitted through sap, 32 aphid species are involved in transmission.

Favorable conditions

» Temperature around 18°C and humid weather.

Management

» Deep summer ploughing.

» Use resistant or tolerant varieties.

» Use healthy/ certified seeds and keep the field free from weeds.

» Rogue out infected plants and burn them

» Pre-sowing soil application of phorate at 10 kg/ ha.

» Sprays thiamethoxam 25 WG at 100 g/ ha or methyl demeton 800 ml/ ha at 30 and 45 days after sowing.

Chapter - 20

Diseases of Moth bean
- *Vigna aconitifolia* (Jacq.) Marechal

S. No.	Disease	Pathogen
1.	Wilt	*Fusarium oxysporum*
2.	Charcoal rot	*Macrophomina phaseolina*
3.	*Cercospora* leaf blight	*Cercospora* spp.
4.	Bacterial blight	*Pseudomonas syringae* pv. *phaseolicola*
5.	Bacterial leaf spot	*Xanthomonas phaseoli*
6.	Mungbean yellow mosaic disease	*Mungbean yellow mosaic virus*

1. Wilt

Symptoms

» The first symptom of the disease in the field is drooping of the plants followed by sudden death.

» The leaves may also turn yellow and drop off prematurely.

» Pod formation is severely affected. In collar regions of the wilted plants, necrosis and discolouration can be seen.

» The diseased plants can be pulled out easily than the healthy ones.

- » When the diseased stem is cut, there is a dark brown, discoloured band around the vascular system.

- » Infection occurs directly through the root hairs.

Fungus: *Fusarium oxysporum* (Schlecht) Snyder & Hansen

Disease cycle

- » The fungus survives for indefinite periods in the soil.

- » This fungal disease may spread through infected plant debris and in seeds.

Favourable conditions

- » It is more severe during hot, dry weather conditions and particularly when plants are under stress.

Management

- » Soil amendment with organic manure or enriched coirpith.

2. Charcoal rot / Ashy stem blight

Symptoms

- » The disease symptom starts as yellowing of lower leaves, followed by drooping and defoliation.

- » The stem portion near the ground level shows dark brown lesions and bark at the collar region shows shredding.

- » The sudden death of plants is seen in patches. In the grown-up plants, the stem portion near the soil level shows large number of black pycnidia.

- » The stem portion can be easily pulled out leaving the rotten root portion in the soil.

- » The infection when spreads to pods, they open prematurely and immature seeds shriveled and become black in colour.

- » Minute pycnidia are also seen on the infected pods and seeds.

- » The rotten root as well as stem tissues contains a large number of minute black sclerotia on the infected pods and seeds.

Diseases of Moth bean

Fungus: *Macrophomina phaseolina* (Tassi) Goid.

Disease cycle

» *M. phaseolina* survives as microsclerotia in the soil and on infected plant debris. The microsclerotia serve as the primary source of inoculum and have been found to persist within the soil up to three years.

» Seeds may also carry the fungus in the seed coat.

Favourable conditions

» Germination of the microsclerotia happen mostly temperature between 28°C-35°C.

» High soil temperature and low soil moisture.

Management

» All the infected plants should be removed carefully and destroyed.

» Follow intercropping cropping system (Moth bean: Sesame) (1:1 ratio).

» Irrigate field every two weeks to avoid stress contitions

» Seed treatment with carbendazin 2 g/ kg or captan at 3 g /kg or thiophanate methyl at 2 g/ kg.

» The fields may be irrigated when soils dry up and temperature rises.

3. *Cercospora* **leaf blight**

Symptoms

» On infected leaves (especially those more mature) look for brown or rust-coloured lesions that vary from circular to angular, are 2-10 mm, and may coalesce.

» Lesions may have a grey centre with a slightly reddish border.

» Severely affected leaves become chlorotic. Lesions may dry and portions may fall out, giving the leaf a shot-hole appearance.

» Lesions and blemishes may occur on branches, stems and pods.

Fungus: *Cercospora* spp.

Disease cycle

» The fungus survives in the infected plant debris and in seeds.

Favourable conditions

» Rain and damp weather favour disease development.

» Leaf spot occurs primarily when temperature are optimum and moist weather.

Management

» Seed treatment with carbendazim 2g /kg.

» Spraying the crop with mancozeb 75 WP at 2.5 g/litre

4. Bacterial blight

Symptoms

» Halo blight occurs primarily when temperatures are cool.

» Light greenish-yellow circles that look like halos form around a brown spot or lesion on the plant. With age, the lesions may join together as the leaf turns yellow and slowly dies.

» Stem lesions appear as long, reddish spots. Leaves infected with common blight turn brown and drop quickly from the plant.

Bacterium: *Pseudomonas syringae* pv. *phaseolicola* Van Hall

Disease cycle

» The bacterium survives in the infected plant debris and in seeds.

» The secondary spread is by rain water.

Favourable conditions

» Rain and damp weather favor disease development.

» Blight occurs primarily when temperatures are cool and moist weather.

Diseases of Moth bean

Management

> Seed treatment with streptocyline 0.01% + captan 2 g /kg

5. Bacterial leaf spot

Symptoms

> Many small, large and irregular brown necrotic spots appear on the leaves proving very severe on the leaves.

> These patches are more prominent on upper than on lower leaf surface.

> Minute water-soaked round irregular spots appear in group on leaf surface, their sizes increase and tum brown to black.

> In the extreme cases, leafmay fall down.

> Ultimately petioles, stems and pods may show extended brown spots.

Bacterium: *Xanthomonas phaseoli* (Burkholder) Starr & Burkholder.

Management

> Grow resistant genotypes like IC 8833, Amravati local, PLM 11 RDM 63, RDM 168 and RDM 182.

> Seed treatment with streptocyline 0.01% + captan 2 g /kg

> Spray copper oxychloride 0.3% or streptocycline 0.01% + copper oxychloride 0.3% or carbendazim 0.05% + copper oxychloride 0.3%.

6. Mung bean yellow mosaic disease

Symptoms

> Diseased plant leaves are yellow and small in size.

> Initial visible signs are the appearance of yellow spots scattered over the leaf, expanding rapidly.

> Leaves depict yellow patches alternatively with green areas, the latter may tum yellow.

» Completely yellowish leaves may give whitish look and ultimately may become necrotic.

» Plants may show decreased growth.

» Plants bear few, small and curled pods which bear few and shriveled seeds.

» Plants show distinct stunted growth, golden yellow look of the leaves and their curling behaviour.

Virus: *Mung bean yellow mosaic virus* (MYMV)
Vector: *Bemesia tabaci* (Whitefly)

Management

» Grow resistant varieties like CAZRl Moth-l, Jawala, RMO 40, RMO 257 and RMO 225.

» Grow resistant genotypes PLMO 12, IC 36096, IC 415152, IC 129177, IC 129177, IC 472217, IC 36392, IC 36649, IC 129208, IC 36467, IC 129194, PLMO 30, IC 36573 and RMO 40.

» Grow guar as trap crop for checking white flies.

Chapter - 21

Diseases of Kidney bean / French bean - *Phaseolus vulgaris* L.

S. No.	Disease	Pathogen
1.	Anthracnose	*Colletotrichum lindemuthianum*
2.	Rust	*Uromyces fabae*
3.	Leaf spot	*Cercospora malayensis*
4.	Powdery mildew	*Erysiphe polygoni*
5.	Common bean mosaic	*Bean common mosaic virus*

1.   Anthracnose

Symptoms

» Black sunken spots with reddish or yellow margins on all portion above ground lesions on hypocotyls cause death.

» Seeds inside the pod also infested.

Fungus: *Colletotrichum lindemuthianum* (Sacc. & Magnus) Briosi & Cavara, (1889)

Management

» Seed treatment with carbendazim 1 g or mancozeb at 3 g/ kg.

208   Diseases of Field Crops and their Management

» Spray with carbendazim at 1 g/ lit or mancozeb 2 g/ lit.

» Remove affected parts and burn.

2. Rust

Symptoms

» The yellow spots in cluster are earlier symptoms.

» Later no they convert in dark brown to black longitudinal lesions.

» In severe case entire plants may be killed.

Fungus: *Uromyces fabae* (Pers.) J. Schröt.

Management

» Seed treatment with carbendazim 1 g or mancozeb at 3g/ kg.

» Spray the crop with mancozeb 2 g/ lit.

3. Leaf spot

Symptoms

» Small circular or irregular spot

Fungi: *Cercospora malayensis* F. Stevens & Solheim; *Cercospora abelmoschi* S. Narayan, Kharwar, R.K. Singh & Bhartiya

Management

» Seed treatment with carbendazim 1 g or mancozeb at 3 g/ kg.

» Spray with carbendazim at 1 g/ lit or mancozeb 2 g/ lit.

» Remove affected parts and burn

4. Powdery mildew

Symptoms

» White powdery spots develop on both sides of the leaves and other parts of the plant except roots.

Diseases of Kidney bean / French bean

» In severe infection defoliation occurs.

» Diseased plant debris should be collected and destroyed.

Fungus: *Erysiphe polygoni* DC.

Management

» Spray with wettable sulphur or carbendazim at 0.1%.

5. Common bean mosaic

Symptoms

» Severe mosaic molting with slight puckering and swelling of leaves and stipules, plants stunted and produce fewer pods.

Virus: *Bean common mosaic virus*

Management

» Rogue out infested plants and burn them.

» Select disease free seed.

» Adopt crop rotation if infestation is severe.

Chapter - 22

Diseases of Cowpea -
Vigna unguiculata (L.) Walp.

S. No.	Disease	Pathogen
1.	Wilt	*Fusarium oxysporum* f. sp. *tracheiphilum*
2.	Dry root rot and stem blight	*Macrophomina phaseolina*
3.	*Cercospora* leaf spot	*Cercospora canescens*
4.	Anthracnose and die back	*Colletotrichum lindemuthianum*
5.	Powdery mildew	*Erysiphe polygoni*
6.	Brown rust	*Uromyces phaseoli* var. *typica*
7.	Southern blight	*Sclerotium rolfsii*
8.	Bacterial blight	*Xanthomonas axonopodis* pv. *vignicola*
9.	Cowpea aphid-borne mosaic	*Cowpea aphid-borne mosaic virus*

1. Wilt

Symptoms

» Sudden wilting of whole young plants.

» In mature plants wilt is preceded by stunting of plant yellowing of leaves, withering and falling of leaves leaving base stem.

Diseases of Cowpea

» Roots and lower stem portion rot leading to drying up of the plant.

» Vascular system is discolouration brownish purple.

Fungus: *Fusarium oxysporum* f. sp. *tracheiphilum*

Infects soybean also microconidia are hyaline oblong or oval and single celled. Macroconidia are falcate-shaped, hyaline, thin walled and 4-5 septate. Chlamydospores present.

Disease cycle

In infected plant debris, seed-borne, irrigation water and chlamydospores in soil.

Favourable conditions

» Temperature of 20-25°C and moist humid weather.

» Presence of root-knot nematode *Meloidogyne javanica*.

Management

» Remove and destroy infected plants with roots.

» Grow resistant varieties.

» Adopt crop rotation excluding soybean.

» Treat seed with carbendazim 2 g/ kg or *Trichoderma viride* 4 g/ kg and soil drench with carbendazim 0.1%.

2.   Dry root rot and stem blight / Charcoal rot

Symptoms

» Greyish black, sunken lesions appear on the lower stem which extends up the stem and down the roots.

» Elongated lesions (cankers) are with grey centre with brown margins.

» Canker girdles the stem and eventually the stem breaks.

» Wilting of plants due to vascular plugging.

» Infestation of root knot nematodes increases the disease incidence.

212 Diseases of Field Crops and their Management

Fungus: *Macrophomina phaseolina* (Tassi) Goid.

It has wide host range. Pycnidia are ostiolate and produces hyaline, cylindrical to ovate, single celled pycnidiospores. Sclerotia are minute and black.

Disease cycle

In infected plant debris in soil, sclerotia in soil, wind-borne pycnidiospores, and sclerotia in irrigation water and farm implements.

Favourable conditions

- » Excess soil moisture and presence of root-knot nematode in soil.

Management

- » Plant debris free seeds.
- » Treat seeds with carbendazim 2 g/ kg or *Trichoderma viride* 4 g/ kg.
- » Remove diseased plant with roots and destroy.
- » Crop rotation with rice and grow resistant varieties

3. *Cercospora* leaf spot

Symptoms

Cercospora canescens: Circular, cherry red to reddish brown lesions of 8-15 mm dia on leaf surface. They become silvery grey due to sporulation when spots are numerous, leaves turn yellow and fall.

Cercospora cruenta: Spots are circular to more elongated, 5-11 mm in dia, first visble on lower leaf surface. Spots are reddish brown or purple but become grey due to greyish black mat of mycelium and conidia. They are often with short holes. Stems and pods are also attacked.

Fungi: *Cercosporiopsis canescens* (Ellis & G. Martin) Miura;
       Cercospora canescens Ellis & G. Martin; and *C. cruenta.*
       Teleomorph: *Mycosphaerella cruenta* Ratham
       (Syn: *Pseudocercospora cruenta* (Sacc.) Deighton)

Diseases of Cowpea

Cercosporiopsis canescens: Conidiophores are pale to medium brown elongate with 2 or more prominent scars. Conidia are hyaline, 5-15 septate, smooth, a circular to obclavate and straight or curved.

Cercospora cruenta: Conidiophores are straight and longer than *C.canescens.* Conidia are sub-hyaline, obclavate to cylindrical and multiseptate.

Disease cycle

C. cruenta infects velvet bean, lime bean, French bean, hyacinth bean, long bean (*Vigna unguiculata* sub.sp. *sequeinpedalis*) and Bambara grounds. Infected plant debris (IPS) is soil, collateral hosts and infected seeds, dormant mycelium in infected plant debris, wind-borne conidia.

Favourable conditions

- » Prolonged high humidity for 4 days.

Management

- » Sow disease free certified seeds.
- » Collect and destroy diseased fallen leaves.
- » Grow resistant varieties.
- » Avoid closer spacing.
- » Intercropping with maize or sorghum in alternate rows of cowpea.
- » Plant new crops far from old diseased field.
- » Destroy volunteer plants before new planting
- » Spray mancozeb 0.2% or carbendazim 0.1%.

4. Anthracnose and die back

Symptoms

- » It is chief disease of stem.
- » Lesions are sunken, tan to brown colour and elongate in die curve, coalesce and cause girdling of stem.

» Spotting on the leaves with light brown centre and reddish brown margin severe spotting causes leaf drying and premature leaf fall.

» In severe cases lesions on branches, pedumucles petioles and pods are found.

» Minute black fruiting bodies (acervuli) are seen especially on stems and branches and on leaf spots.

Fungus: *Colletotrichum lindemuthianum* (Sacc. & Magnus)

Acervuli contain conidiophores, conidia and setae. Conidiophores are short, erect, hyaline and umbranched. Conidia are hyaline, single celled and oblong or cylindrical with rounded ends.

Disease cycle

» Infected plant debris, seeds, wind-borne conidia and wind-driven rains.

Favourable conditions: Cool and wet weather

Management

» Destroy diseased plant debris and follow 3 year crop rotation

» Sow disease free certified seeds.

» Mixed cropping and early planting.

» Spray mancozeb 0.2%. or carbendazim 0.1%

5. Powdery mildew

Symptoms

» Small, white powdery patches on the upper leaf surfaces.

» With advancement of the disease, the fungal patches increase in size and cover entire leaf surface.

» Petioles, young stems and pods are also covered with white powdery coating.

» White powdery growth turns brown with small black sexual fruiting bodies called cleistothecia and causes drying and falling of leaves.

Diseases of Cowpea

» Flowering and pod formation and seed setting are reduced.

» Seeds are shriveled and small in size with poor germination.

Fungus: *Erysiphe polygoni* DC

Obligate parasite. Conidiophores are septate and arise vertically from superficial hyaline, septate mycelium. They bear conidia in chain. Conidia are barrel-shaped, hyaline and single celled.

Disease cycle

» Dormant mycelium in diseased plant debris and volunteer cowpea plant and other volunteer plants

» Sspread by wind-borne conidia.

Favourable conditions: Warm humid weather and shady areas.

Management

» Grow resistant varieties.

» Spray carbendazim or wettable sulphur 0.3%.

6. Brown rust

Symptoms

» Light brown uredia appear on both the surfaces releasing reddish brown powdery uredospores.

» Uredia are surrounded by yellow haloes.

» Intense spotting with uredia causes the leaves to wilt, droop and dry prematurely.

Fungus: *Uromyces phaseoli* var. *typica* Arthur.

Uredosori are brown. Uredospores are ovate, light brown and spiny. Telia are black-teliospores are ovate, chestnut brown with broad hyaline papilla at the tip.

Disease cycle

» As teliospores in diseased plant debris, from volunteer plants

» Spread by wind-borne uredospores.

Favourable conditions

» Cloudy humid weather with heavy dew and temperature of 22-28°C.

Management

» Follow crop rotation.

» Plant the seeds early in the season.

» Spray carbendazim 0.1% or wettable sulphur 0.3%.

7.   Southern blight

Symptoms

» The fungus attacks the basal stem portion and fan or silky white mycelium appears mustard seed-like brown sclerotia are seen.

» Diseased plant sudden wilt and foliage becomes yellow and the plants die.

» Diseased plants can be easily hand pulled.

Fungus: *Sclerotium rolfsii* Sacc.

Larger mustard seed-like brown sclerotia are produced.

Disease cycle

» Diseased plant debris in soil, weed hosts and externally seed-borne.

» Sclerotia by wind-blown soil.

» Irrigation wate and by movement of soil by farm implements and workers.

Favourable conditions

» Overcrowding of plants.

Diseases of Cowpea

» High temperature and high relative humidity and acdic soil.

Management

» Avoid overcrowding of plants to promote air circulation.

» Rotate with non-host crops.

» Deep summer ploughing and soil solarization.

8.   Bacterial blight / Bacterial canker

Symptoms

» Seedling from infected seed dies.

» Cotyledons turn red and shrivel and a necrotic spot on primary leaves spread to growing point leading to death of seedling.

» Surviving seedlings develop canker on the stem at the point of attachment of cotyledons.

» Primary leaves, enlarges and turns brown and stem breaks at cankered region.

» On leaves of matured plants light yellow round to irregular spots of 4-10 mm appear.

» The spots are with necrotic brown centre and deep yellow margin.

» Veins necrosis and yellowing of adjoining areas cause the leaves to become straw coloured and to defoliate prematurely.

» Dark green water-soaked streaks along the structures of pods lead to yellowing, shriveling and drying of pods.

» Seeds are small and shriveled with poor germination.

Bacterium: *Xanthomonas axonopodis* pv. *vignicola* (Burkholder) (Syn:: *X. campestris* pv. *vignicola*) (*X. vignicola*). Infected sunnhemp, lablab, French bean, black nightshade, urd bean and *Tephrodia purpurea*.

Disease cycle

» Diseased plant debris and infected seeds.

»   Infected seed.

»   Wind-driven rain, insect and wind-blown soil.

Favourable conditions: Temperature of (>25°C) and relative humidity (>75%), overhead irrigation and heavy rainfall.

Management

»   Plant only disease free certified seeds.

»   Grow resistant varieties.

»   Follow strict plant quarantine.

»   Intercropping cowpea with maize or castor.

»   Immerse seeds in hot water at 50°C for 10 min.

9.   Cowpea aphid-borne mosaic

Symptoms

»   Irregular, chlorotic areas on leaves lead to mosaic mottling on young trifoliate leaves.

»   Veins banding, puckering and distortion of leaves.

»   Severe infection causes reduction in leaflet size and blistering and bleaching of leaves.

»   Pods are curved, twisted and become small

»   Seeds lesser in number per pod and shriveled.

Virus: *Cowpea aphid-borne mosaic virus* (CABMV).

»   Flexuous filamentous particles of 750 nm.

Transmission: Aphids- *Aphis craccivora, A. fabae, A. gossypii* and *Macrosiphon enphorliae.* Not transmitted through sap.

Disease cycle: Survives on weed hosts, sunnhemp and French bean seed borne.

Favourable conditions: Abundant aphid vector population.

Management

- » Grow resistant varieties.
- » Remove and destroy diseased plant periodically.
- » Spray methyl dematon 0.05% to control insect vectors.

Minor diseases

Bacterial blight	*Xanthomonas phaseoli*	Refer Black gram
Seedling blight	*Sclerotium rolfsii*	Refer Red gram
Stem rot	*Pythium aphanidermatum*	Refer Red gram
Cottony rot	*Sclerotinia sclerotiorum*	Refer Bengal gram
Stem rot	*Phytophthora cactorum*	Dark brown or blackbrown sunken lesionsappear at the base ofstem or branches, extending several centimeters. Theleaves are light green with upward rolling andrapid wilting occurs.
Red leaf spot	*Septoria vignae*	Dark red, circular orirregular spots appear on both the surfaces of leaves.
Brown leaf spot	*Myrothecium roridum*	Small brown spotswith pinkish margin appear on leaves. Lesions may alsooccur on petiole, stem and pod.
Die-back	*Colletorichum capsici*	Drying of twigs from the tip to downwards
Root rot	*Pellicularia koleroga*	Disintegration and rotting of root system

Chapter - 23

Diseases of Lablab - *Lablab purpureus* (L.) Sweet

S. No.	Disease	Pathogen
1.	Anthracnose	*Colletotrichum lindemuthianum*
2.	Powdery mildew	*Erysiphe polygoni*
3.	Rust	*Uromyces fabae*
4.	Root rot	*Fusarium solani* f. sp. *phaseoli*
5.	Stem rot	*Sclerotinia sclerotiorum*
6.	Bacterial blight	*Xanthomonas campestris* pv. *phaseoli*
7.	Mosaic	*Bean yellow mosaic virus*

1.   Anthracnose

Symptoms

> » Lesions on stems and pods more clearly defined than those on leaves, grey or brown, slightly sunken with raised dark brown or reddish edge.

> » All vegetative parts, except pulvini, are susceptible during early stages of development; invasion of the tap root of a young plant can lead to death.

> » Elongated dark-brown or black sunken spots with reddish or yellowish margins appear on veins, petioles, stem and pods.

Diseases of Lablab

» Diseased seeds carry the fungus from season to season.

» Spots on the hypocotyl cause death of the plant.

» Seedlings show canker on cotyledons.

» These lesions produce pinkish spore masses during moist weather.

Fungi: *Colletotrichum lindemuthianum* (Sacc. & Magnus) and *Colletotrichum capsici* (Syd).

Management

» Sowing disease-free seed, follow crop rotation and field sanitation.

» Treat seeds for half an hour in solution of Tolyl mercury acetate (ceresan 0.125%) or organomercurial fungicide (Seedex 1%).

» Dust seedex or captan at 3 g/ kg.

» Spray Bordeaux mixture 5:5:50 or a copper fungicide at 1 kg in 250 litres of water.

2. Powdery mildew

Symptoms

» Lablab bean is affected mostly by *Leveillula taurica* var. *macrospora*.

» White floury patches are formed on both sides of the leaves and all portions of the above ground parts, which gradually turn brown.

» The leaves become yellow and die, while fruits either do not set or remain small.

» It is a seed borne as well as a soil-borne disease and causes serious damage in the dry weather.

Fungi: *Erysiphe polygoni* DC.; *Leveillula taurica* Lev.

Management

» To control this disease, field sanitation is important.

» Diseased plants should be collected and burnt.

» The crop may also be dusted with 200-mesh sulphur at the rate of 25-30 kg per hectare at 10-15 days interval or spray wettable sulphur at 1 kg in 500 liters water.

3. Rust

Symptoms

» Red and black pustules appear on the lower surface of the leaves.

» Attacked leaves turn yellow and drop off.

Fungi: *Uromyces fabae* de Bary ex Cooke, *U. appendiculatus* and *U. mucunae.*

Management

» Dusting with sulphur at the rate of 25-30 kg/ ha protects the crop.

» Spraying Bordeaux mixture in the early stages also prevents spread.

4. Root rot

Symptoms

» Lesions caused by *Fusarium* are elongate and red brown.

» This is a soil-borne fungus favoured by cold wet soil conditions.

Fungus: *Fusarium solani* f. sp. *phaseoli*

Management

» Resistant varieties grown on well drained soil with good tilth will reduce disease levels along with 3-4 year crop rotations.

5. Stem rot (White mold)

Symptoms

» *Sclerotinia* causes major crop losses wherever beans are grown.

» All above ground parts can be attacked especially old flowers.

» Most, if not all, infection arises from ascospores.

» Mycelium on old flower parts moves in to infect pods and leaves and extends to adjoining plants.

Diseases of Lablab

Fungus: *Sclerotinia sclerotiorum* (Lib.) de Bary

Management

» The only effective control is fungicide application during flowering to protect the flowers.

» Crops with dense canopies and a field history of this disease benefit most.

6. Bacterial blight and Halo blight

Symptoms

» Common, fuscous and halo blights are frequent and destructive diseases of dry beans.

» All three diseases are difficult to distinguish from one and other in the fields.

Bacterium: *Xanthomonas campestris* pv. *phaseoli, X. phaseoli* var. *fuscans,*

» *Pseudomonas syringae* pv. *phaseolicola*

Management

» Control is via clean seed, avoidance of working fields when the foliage is wet and copper based fungicides.

7. Mosaic: *Bean yellow mosaic virus*

» Bean yellow mosaic is aphid transmitted but is not seed transmitted as is the case with common mosaic.

Chapter - 24

Diseases of Indian Pea -
Lathyrus sativus L

S. No.	Disease	Pathogen
1.	Wilt	*Fusarium oxysporum orthoceros* var. *lathyri*
2.	Rust	*Uromyces viciae-fabae*
3.	Downy mildew	*Peronospora lathyri-palustris*
4.	Powdery mildew	*Leveillula taurica*

1. Wilt

Symptoms

- » Premature wilting of whole young plants.
- » In mature plants wilt is preceded with yellowing of leaves, withering and falling of leaves leaving base stem.
- » Vascular system is discolouration brownish purple.

Fungi: *Fusarium oxysporum orthoceros* var. *lathyri* V.P. Bhide & Uppal; *Fusarium oxysporum* f. sp. *ciceris* (Padwick) Snyder and Hansen

Diseases of Indian pea

Disease cycle

In infected plant debris, seed-borne, irrigation water and chlamydospores in soil.

Favourable conditions

- » Temperature of 20-25°C
- » Moist humid weather

Management

- » Remove and destroy infected plants with roots.
- » Treat seed with carbendazim 2 g/ kg or *Trichoderma viride* 4 g/ kg and soil drenching with carbendazim 0.1%.

2. Rust

Symptoms

- » Pink to brown pustules appeared on leaves and stems.
- » In severe attack, the affected plants any dry.

Fungi: *Uromyces viciae-fabae* (Pers.) Schroe*t*; *Euromyces fabae* de Bary ex Cook

Management

- » Grow early maturing variety.
- » Seed treatment with carbendazim at 2g/ kg seed.
- » Spray the crop with mancozeb 75 WP at 2 g/ litre of water

3. Downy mildew

Symptoms

- » Brownish cottony growth of fungus may be seen on the lower surface of leaf.
- » Inside growth yellow to greenish spots are also visible.

Fungus: *Peronospora lathyri-palustris* Gaumann

Management

> Spray the crop with mancozeb 75 WP at 2 g/ litre of water.

4. Powdery mildew

Symptoms

> Symptoms first appeared on all the aerial part of plant.

> While powdery masses of spores formed on leaves which may collapse and cover the whole leaf with powdery growth.

Fungi: *Leveillula taurica* Lev.; *Erysiphe pisi* DC; *Erysiphe polygoni* DC

Management

> Spray with wettable sulphur at 3 gm/ litre or carbendazim at 1g/ litre.

Chapter - 25

Diseases of Lentil - *Lens culinaris* Medic.

S. No.	Disease	Pathogen
1.	Wilt	*Fusarium oxysporum* f. sp. *lentis*
2.	*Botrytis* grey mold	*Botrytis fabae*
3.	Rust	*Uromyces viciae-fabae*
4.	*Ascochyta* blight	*Ascochyta lentis*
5.	*Stemphylum* blight	*Stemphylium botryosum*

1. Wilt

Symptoms

» The disease appears in the field in patches at both seedling and adult stages.

» Seedling wilt is characterized by sudden drooping, followed by drying of leaves and seedling death.

» The roots appear healthy, with reduced proliferation and nodulation and usually no internal discoloration of the vascular system.

» Adult wilt symptoms appear from flowering to late pod-filling stage and are characterized by sudden drooping of top leaflets of the affected plant, leaflet closure without premature shedding, dull green foliage followed by wilting of the whole

228 Diseases of Field Crops and their Management

plant or individual branches.

» Seeds from plants affected in mid-pod-fill to late pod-fill are often shriveled.

Fungus: *Fusarium oxysporum* f. sp. *lentis* (Vasudeva & Sriniv.) W.L. Gordon

The fungus is a soilborne pathogen, although seed infestation and infection is common. The chlamydospores can survive in soil either in dormant form or saprophytically for several years without a suitable host.

Favourable conditions

» Low soil temperature, 30% soil water holding capacity and increasing plant maturity.

Management

» Soil amendment with organic matter enhances antagonism with other soil microflora

» Ploughing of the field during summer.

» Following crop rotation with cereal crops which are not affected by wilt pathogen.

» Grow resistant varieites like Pant L-4, Pant L-6, Pant L-8 and Noori.

» Using antagonistic microflora like *Bacillus subtilis, Trichoderma harzianum, T. viride* at 4 g/ kg.

» Seed treatment with thiram 0.3%.

» Spray with carbendazim 1 g/ lit or mancozeb 2.5 g/ lit or captan 2.5 g/ lit.

2.   *Botrytis* grey mold

Symptoms

» All above ground plant parts of lentil can be affected by botrytis grey mould.

» The disease first appears on the lower foliage as discrete lesions on leaves which are initially dark green, but turn grayish-brown, then cream as they age, that enlarge and coalesce to infect whole leaflets.

» Severely infected leaves senesce and fall to the ground.

» Lesions girdle the stem and cover it with a furry layer of grey mold, eventually

Diseases of Lentil 229

causing stem and whole plant death.

Fungi: *Botrytis fabae* (Sard) (teleomorph: *Botryotinia fabae*) and *B. cinerea* (Pers.: Fr.)

Disease cycle

» Spread by seed-borne inoculum, sclerotia, mycelium in old infected trash, and alternate host.

» Favourable conditions

» High humidity and moderate temperatures with high moisture favours the diseases

Management

» Grow resistant varieties Pant L-639 and Pant L-406.

» Seed treatments with carboxin 0.1% or chlorothalonil 0.1% or thiabendazole 0.2%.

3.   Rust

Symptoms

» Rust pustules can be seen on leaf blade, petiole and stem.

» Rust starts with the formation of yellowish-white pycnidia and aecial cups on the lower surface of leaflets and on pods, singly or in small groups in a circular form.

» Later, brown uredial pustules emerge on either surface of leaflets, stem and pods.

» Pustules are oval to circular and up to 1 mm in diameter.

» They may coalesce to form larger pustules.

» In severe infections leaves are shed and plants dry prematurely, the affected plant dries without forming any seeds in pods or with small shriveled seeds.

Fungus: *Uromyces viciae-fabae* (Pers.) Schroet

Disease cycle

» It is an autoecious fungus, completing its life cycle on lentil.

» Favourable conditions

» High humidity, cloudy or drizzly weather with temperatures 20 to 22°C

Management

» Grow resistant varieties like Pant L-639, Pant L-406, Pant L-6, pant L-7 and Pant L-8.

» Spray with ferbam or mancozeb 0.2% or triadimefon 0.05 and calixin 0.2%

4. *Ascochyta* blight

Symptoms

» The symptoms of the disease include lesions on leaves, petioles, stems and pods.

» The irregular shaped lesions on leaves, petioles and stem are tan and darker brown on pods and seeds.

» In severe infection, lesions can girdle the stem, leading to breakage and subsequent death of all tissues above the lesion.

» Affected crops under severe infection may be blighted and seed may become shriveled, reduced in size, and discolored.

» Flowers and pods could abort, leading to yield loss.

Fungus: *Ascochyta lentis* Bond. and Vassil

The asexual stage is characterized by the production of pycnidia in the lesions on infected plants, which are 175–300 μm in diameter with a minute round osteole. The pycnidia release conidia which are cylindrical, straight or rarely curved, round at the ends with a median septum.

Disease cycle

» Spread by seed-borne inoculum.

» Favourable conditions

» Disease is favored by cool, moist weather.

» An extended period of leaf wetness is required for disease development

» Management

Diseases of Lentil

» Crop rotation.

» Spray with carbendazim, carboxin, iprodion and thiobendazole at 0.1%.

5. *Stemphylum* blight

Symptoms

» Initially, small pin-headed light brown to tan coloured spots on leaflets.

» The infected tissue appears light cream in colour, often with angular patterns of lighter and darker areas that spread across, or long, the entire leaflet.

» Leaves and stems gradually turn dull yellow, giving a blighted appearance to the crop.

» The infected leaves can be abscised rapidly, leaving only the terminal leaflets on the stems.

» The stems bend down, dry and gradually turn ashy white, but pods remain green.

Fungus: *Stemphylium botryosum* Wallr (Teleomorph: *Pleospora herbarum* (Fr) Rab.).

Disease cycle

» Spread by seed-borne inoculum.

» Favourable conditions

» Disease is favored ambient night temperature remains above 8°C, and the mean day temperature exceeds 22°C.

» The RH inside the crop canopy must also reach 94%.

Management

» Crop rotation.

» Spray with carbendazim at 0.1%.

Chapter - 26

Diseases of Groundnut / Peanut / Monkeynut / Goober pea / Earthnut
Arachis hypogaea L.

S. No.	Disease	Pathogen
1.	Early tikka leaf spot	*Mycosphaerella arachidis*
2.	Late tikka leaf spot	*Mycosphaerella berkeleyii*
3.	Rust	*Puccinia arachidis*
4.	Collar rot	*Aspergillus niger*
5.	Root rot	*Macrophomina phaseolina*
6.	Stem and pod rot	*Sclerotium rolfsii*
7.	Wilt	*Fusarium oxysporum*
8.	Anthracnose	*Colletotrichum dematium*
9.	Yellow mould	*Aspergillus flavus*
10.	Leaf spot	*Alternaria arachidis*
11.	Rossette	*Groundnut rosette virus*
12.	Groundnut bud necrosis disease	*Groundnut bud necrosis virus*
13.	Indian peanut clump disease	*Peanut clump virus*

Diseases of Groundnut

Tikka leaf spots

The disease occurs on all above ground parts of the plant, more severely on the leaves. The leaf symptoms produced by the two pathogens can be easily distinguished by appearance, spot colour and shapes. Both the fungi produce lesions also on petiole, stem and pegs. The lesions caused by both species coalesce as infection develops and severely spotted leaves shed prematurely. The quality and yield of nuts are drastically reduced in severe infections.

1.   Early tikka

Symptoms

» The spots caused by *C. arachidicola* are irregular to circular, larger than those caused by *C. personatum*, surrounded by a bright yellow, circular, halo with a dark brown centre.

Fungus: Teleomorph: *Mycosphaerella arachidis* Deighton (*Cercospora arachidicola* S.Hori)

The pathogen is intercellular and do not produce haustoria and become intracellular when host cells die. The fungus produces abundant sporulation on the upper surface of the leaves. Conidiophores are olivaceous brown or yellowish brown in colour, short, 1 or 2 septate, unbranched and geniculate and arise in clusters. Conidia are sub hyaline or pale yellow, obclavate, often curved 3-12 septate, 35-110 × 2.5-5.4 µm in size with rounded to distinctly truncate base and sub-acute tip. The perfect stage of the fungus produces perithecia as ascostromata. They are globose with papillate ostiole. Asci are cylindrical to clavate and contain 8 ascospores. Ascospores are hyaline, slightly curved and two celled, apical cell larger than the lower cell.

2.   Late tikka

Symptoms

» Intially, small, dark spots which enlarge to about 3-8 mm in diameter.

» A few to several spots may be found on each leaflet. Often they enlarge and coalesce to form irregular dark brown to black, blighted patches.

» They occur mostly on the leaf blade, but may occur also on the petiole and stem.

» Severely infected leaves drop off prematurely.

234 — Diseases of Field Crops and their Management

» The disease usually shows up about a month after sowing and at times it becomes severe, especially between the flowering and harvest period.

» Most of varieties susceptible to late tikka disease is prone to rust disease.

Fungus: Teleomorph: *Mycosphaerella berkeleyii* Jenk.
Anamorph: *Phaeoisariopsis personata* (Berk. & M.A. Curtis) Arx
(Syn: *Cercosporidium personata* (Berk. & M.A. Curtis) Deighton)
(Syn: *Passalora personata* (Berk. & M.A. Curtis)

The fungus produces internal and intercellular mycelium with the production of haustoria. The conidiphores are long, continuous, 1-2 septate, geniculate, arise in clusters and olive brown in colour. The conidia are cylindrical or obclavate, short, measure 18-60 × 6-10 μm, hyaline to olive brown, usually straight or curved slightly with 1-9 septa, not constricted but mostly 3-4 septate. The fungus in its perfect stage produces perithecia as ascostromata which are globose or broadly ovate with papillate ostiole. Asci are cylindrical to ovate, contain 8 ascospores. Ascosporous are 2 celled and constricted at septum and hyaline.

Disease cycle

The pathogen survives for a long period in the infected plant debris through conidia, dormant mycelium and perithecia in soil. The volunteer groundnut plants also harbour the pathogen. The primary infection is by ascospores or conidia from infected plant debris or infected seeds. The secondary spread is by wind blown conidia. Rain splash also helps in the spread of conidia.

Favourable conditions

» Prolonged high relative humidity for 3 days.

» Low temperature (20°C) with dew on leaf surface.

» Heavy doses of nitrogen and phosphorus fertilizers

» Deficiency of magnesium in soil.

Management

» Remove and destroy the infected plant debris.

» Eradicate the volunteer groundnut plants.

Diseases of Groundnut

» Keep weeds under control.

» Adjust the date of sowing to avoid susceptible season.

» Grow early tikka resistant varieties like PI 109839, PI 162857, NC 5 and NC 3033.

» Grow late tikka resistant varieties like PI 261893, PI 262090, PI 371521 and ALR 1.

» Seed treatment with carbendazim at 2 g/ kg.

» Spraying of *Trichoderma viride* 5% and *Verticillium lecanii* 5% can reduce the disease severity.

» Spray sulphur 40 WP 3 kg/ ha or sulphur 80 WP 3 kg/ ha or carbendazim 500 g or tebuconazole 25.9% EC 0.75 lit/ ha or bitertanol 25% WP 2 g/ lit or metiram 70% WG 4 g/ lit or mancozeb 2 kg or chlorothalonil 2 kg/ ha and if necessary, repeat after 15 days.

Difference between early and late tikka leaf spot

Early leaf spot	Late leaf spot
Infection starts about a month after sowing.	Infection starts around 5 – 7 week after sowing
Circular or irregular reddish brown lesion with yellow halo	Small circular dark brown lesion without yellow halo
On lower surface lesion light brown colour. Conidia whip like pale yellow 3-12 septum	Lower surface lesion carbon black colour Conidia obclavate 4-12 septa
Both produces lesions also appear on petioles, stems, pegs. The lesion coalesce and premature dropping of leaves. The quality and yield reduced	

3. Rust

Symptoms

» The disease attacks all aerial parts of the plant.

» The disease is usually found when the plants are about 6 weeks old.

» Small brown to chestnut dusty pustules (uredosori) appear on the lower surface of leaves.

» The epidermis ruptures and exposes a powdery mass of uredospores.

» Corresponding to the sori, small, necrotic, brown spots appear on the upper surface of leaves.

» The rust pustules may be seen on petioles and stem.

» Late in the season, brown teliosori, as dark pustules, appear among the necrotic patches.

» In severe infection lower leaves dry and drop prematurely.

» The severe infection leads to production of small and shriveled seeds.

Fungus: *Puccinia arachidis* Speg.

The pathogen produces both uredial and telial stages. Uredial stages are produced abundant in groundnut and production of telia is limited. Uredospores are pedicellate, unicellular, yellow, oval or round and echinulated with 2 or 3 germpores. Teliospores are dark brown with two cells. Pycnial and aecial stages have not been recorded and there is no information available about the role of alternate host.

Disease cycle

The pathogen survives as uredospores on volunter groundnut plants. The fungus also survives in infected plant debris in soil. The spread is mainly through wind borne inoculum of uredospores. The uredospores also spread as contamination of seeds and pods. Rainsplash and implements also help in dissemination. The fungus also survives on the collateral hosts like *Arachis marginata, A. nambyquarae* and *A. prostrate.*

Favourable conditions

» High relative humidity (> 85%) and heavy rainfall.

» Low temperature (20-25°C).

Management

» Avoid monoculturing of groundnut.

» Remove volunteer groundnut plants and reservoir hosts.

» Grow resistant varieties like ICG 4, ICG 10, Tifrust 12 and Tifrust 13.

» Grow moderately resistant varieties like ALR 1.

» Spray mancozeb 2 kg or Wettable Sulphur 80 WP 3 kg or Difenoconazole 25% EC 0.1% or Chlorothalonil 2 kg/ ha or Tebuconazole 25.9% EC 0.75 lit/ ha or bitertanol 25% WP 2 g/ lit.

Diseases of Groundnut

4. Collar rot / seedling blight / crown rot

Symptoms

» The disease usually appears in three phases.

i. Pre-emergence rot

» Seeds are attacked by soil-borne conidia and caused rotting of seeds.

» The seeds are covered with black masses of spores and internal tissues of seed become soft and watery.

ii. Post-emergence rot

» The pathogen attacks the emerging young seedling and cause circular brown spots on the cotyledons.

» The symptom spreads later to the hypocotyl and stem.

» Brown discolored spots appear on collar region.

» The affected portion become soft and rotten, resulting in the collapse of the seedling.

» The collar region is covered by profuse growth of fungus and conidia and affected stem also show shredding symptom.

iii. Crown rot

» The infection when occurs in adult plants show crown rot symptoms.

» Large lesions develop on the stem below the soil and spread upwards along the branches causing drooping of leaves and wilting of plant.

Fungus: *Aspergillus niger* Tiegh and *A. pulverulentum* Porres; *Sporotrichum pulverulentum* Novobr.,

The mycelium of the fungus is hyaline to sub-hyaline. Conidiophores arise directly from the substrate and are septate, thick walled, hyaline or olive brown in colour. The vesicles are mostly globose and have two rows of hyaline phialides *viz.*, primary and secondary phialides. The conidial head are dark brown to black. The conidia are globose, dark brown in colour and produce in long chains.

Disease cycle

»   The pathogen survives in plant debris in the soil, not necessarily from a groundnut crop.

»   Soil-borne conidia cause disease carry over from season to season.

»   The other primary source is the infeced seeds.

»   The pathogen is also seedborne in nature.

Favourable conditions

»   Deep sowing of seeds.

»   High soil temperature 30-35°C and low soil moisture.

Management

»   Crop rotation with non-host crops.

»   Destruction of infected plant debris.

»   Resistant varieties like EC 21115, J 11, JCG 88 and OG-52-1.

»   Remove and destroy previous season's infested crop debris in the field

»   Seed treatment with *Trichoderma asperellum / T. harzianum* at 4 g/ kg or furrow application at 148 lit solution (1.38×10^3 CFU) in 370 kg wheat husk or furrow application at 1.5 kg/ ha (1×10^6 CFU) by mixing with soil.

Disease	Pathogen	Bioagent
Collar rot	*Aspergillus niger*	*Trichoderma harzianum*
Seed rot	*Aspergillus flavus*	*Trichoderma viride*
Crown rot	*Sclerotium rolfsii*	*Trichoderma virens*
Web blight	*Rhizoctonia solani*	*Bacillus subtilis*

»   Soil application of tebuconazole 2% DS 0.25% or *Trichoderma asperellum / T. harzianum* at 2.5 kg/ ha, preferably with organic amendments such as castor cake or neem cake or mustard cake at 500 kg/ ha.

»   Spray with carbendazim 25% + mancozeb 50% WS 0.1%.

Diseases of Groundnut

5. Root rot

Symptoms

» In the early stages of infection, reddish brown lesion appears on the stem just above the soil level.

» The leaves and branches show drooping, leading to death of the whole plant.

» The decaying stems are covered with whitish mycelial growth.

» The death of the plant results in shredding of bark.

» The rotten tissues contain large number of black or dark brown, thick walled sclerotia.

» When infection spreads to underground roots, the sclerotia are formed externally as well as internally in the rotten tissue.

» Pod infection leads to blackening of the shells and sclerotia can be seen inside the shells.

Fungus: *Macrophomina phaseolina* (Tassi) Goidanich
Rhizoctonia bataticola (Tassi) E.J. Butler

The fungus produces hyaline to dull brown mycelium. The sclerotia are thick walled and dark brown in colour.

Disease cycle

The fungus remains dormant as sclerotia for a long period in the soil and in infected plant debris. The primary infection is through soil-borne and seed-borne sclerotia. The secondary spread of sclerotia is aided by irrigation water, human agency, implements and cattle etc.

Favourable conditions

» Prolonged rainy season at seedling stage and low lying areas.

Management

» Seed treatment with carbendazim 2 g/ kg or tebuconazole 0.25%.

» Seed treatment with biological agent *Trichoderma asperellum* at 4 g/ kg.

240   Diseases of Field Crops and their Management

» Soil application by mixing 1 kg of *Trichoderma* in 25 kg of FYM, allow growing the bioagent. Mix 25 kg of *Trichoderma* grown FYM in 100 kg of fresh FYM and spread over (1 ha) and mix well.

» Spot drench with carbendazim at 0.5 g/ lit.

6. Stem and pod rot

Symptoms

» The first symptom is the sudden drying of a branch which is completely or partially in contact with the soil.

» The leaves turn brown and dry but remain attached to the plant.

» Near soil on stems white growth of fungus mycelium is appeared.

» As the disease advances white mycelium web spreads over the soil and the basal canopy of the plant.

» The sclerotia, the size and colour of mustard seeds, appear on the infected areas as the disease develops and spreads.

» The entire plant may be killed or only two or three branches may be affected.

» Lesions on the developing pegs can retard pod development.

» Infected pods are usually rotted.

Fungus: *Sclerotium rolfsii* Sacc.

Management

» Cultural practices such as deep' covering or burial of organic matter before planting, non-dirting cultivation by avoiding movement of soil up around the base of plants and preventing accumulation of organic debris are extremely useful in reducing the disease.

» Crop rotation with wheat, corn and soyabean may minimize the incidence of stem rot.

» Grow resistant varieties like ICGV 86416, ICGV 87359, NC 9 and GAT 141.

» Seed treatment with carbendazim or captan at 2-3 g/ kg seed.

Diseases of Groundnut

» Seed treatment with *Trichoderma asperellum* formulation (4 g/ kg) followed by application of 2.5 kg *Trichoderma asperellum* formulation mixed with 50 kg farm yard manure before sowing.

7. Wilt

Symptoms

» Germinating seeds are attacked by the pathogens shortly before emergence.

» There is general tissue disintegration and the surface of the seedling is covered with sporulating mycelium.

» Damping off symptoms characterized by brown to dark brown.

» Water soaked sunken lesions on the hypocotyl which later encircle the stem and extend above the soil level.

» Roots are also attacked, especially the apical portions.

» The affected seedlings become yellow and wilted.

» The leaves turn greyish green and the plants dry up and die.

» The roots and stems show internal vascular browning and discolouration.

» These fungi are also commonly associated with pod rot.

Fungi: *Fusarium oxysporum* Schlecht. emend. Snyder & Hansen and *F. solani* (Mart.) Sacc.

Management

» Seed treatment with carbendazim at 2 g/ kg seed.

8. Anthracnose

Symptoms

» Small water-soaked yellowish spots appear on the lower leaves which later turn into circular brown lesions with yellow margin 1 to 3 mm in diameter.

» In some cases lesions enlarge rapidly become irregular and cover the entire leaflet, and extend to the stipules and stems.

242   Diseases of Field Crops and their Management

»   Brownish grey lesions occur on both the surfaces of leaflets.

»   Infection spreads to stipules, petioles and branches.

Fungi: *Colletotrichum dematium* (Pers.) Grove and *Colletotrichum capsici* (Sydow)

Disease cycle

»   The pathogen is seed, soil and air-borne.

Management

»   Deep summer ploughing.

»   Use healthy certified seeds and removal of plant debris.

»   Seed treatment with copper oxychloride at 3 g/ kg or carbendazim at 2 g/ kg.

9.   Yellow mould

Symptoms

»   Seed and un-emerged seedlings attacked by the pathogen are rapidly shriveled and dried.

»   Brown or black mass covered by yellow or greenish spores may be seen.

»   Decay is most rapid when infected seeds are planted. After seedling emergence cotyledons already infected with the pathogen, show necrotic lesions with reddish brown margins.

»   This necrosis terminates at or near the cotyledonary axis. Under field conditions the diseased plants are stunted, and are often chlorotic.

»   The leaflets are reduced in size with pointed tips, widely varied in shape and sometimes with veinal clearing.

Fungus: *Aspergillus flavus* Link.

Management

»   Rapid germination and vigorous growth of seedlings will reduce the infection.

Diseases of Groundnut

» Seed treatment with carbendazim or captan at 2 g/ kg seed.

10. Leaf spot

Symptoms

» Lesions produced by *A. arachidis* are brown in colour and irregular in shape surrounded by yellowish halos.

» Lesions produced by *A. tenuissima* are characterized by blighting of apical portions of leaflets which turn light to dark brown colour.

» Lesions produced by *A. alternata* are small, chlorotic, water soaked, that spread over the surface of the leaf.

» The lesions become necrotic and brown and are round to irregular in shape.

» Veins and veinlets adjacent to the lesions become necrotic.

» Lesions increase in area and their central portions become pale, rapidly dry out, and disintegrate.

» Affected leaves show chlorosis and in severe attacks become prematurely senescent.

» Lesions can coalesce; give the leaf a ragged and blighted appearance.

Fungi: *Alternaria arachidis* R.L. Kulk and *Alternaria tenuissima* Samuel Paul Wiltshire.

Management

» Spray mancozeb 2 kg/ ha or copper oxychloride 2 kg/ ha or carbendazim 500 g/ ha.

11. Rossette

Symptoms

» The affected plants are characterized by the appearance of dense clump or dwarf shoots with tuft of small leaves forming in a rosette fashion.

» The plant exhibits chlorosis and mosaic mottling.

» The infected plants remain stunted and produce flowers, but only a few of the pegs may develop further to nuts but no seed formation.

Virus: The disease is caused by a complex mixture of viruses *viz.,Groundnut rosette assistor virus* (GRAV), *Groundnut rosette virus* and *Groundnut rosette satellites* is an isometric, not enveloped and 28 nm diameter (reported from India) and it gives no overt symptom in groundnut. *Groundnut rosette virus* is with ssRNA genome, which becomes packaged in GRAV virious and thus depends on it for aphid transmission, but produces no overt symptoms in groundnut. The groundnut rosette satellites are satellite RNAs that control the symptoms and cause the different types of rosette (chlorotic, green and mosaic).

Disease cycle

The primary source of spread by aphid vector, *Aphis craccivora* and *A. gossipii* in a persistent manner, retained by vector but not transmitted congenitally. The virus is not transmitted by any other means like mechanical or seed or pollen. The virus can survive on the volunteer plants of groundnut and other weed hosts.

Management

- » Practice clean cultivation, use heavy seed rate and rogue out the infected plants.
- » Grow resistant varieties like RMP 12 and RMP 91.
- » Spray monocrotophos or methyl demeton at 500 ml/ ha.

12. Groundnut bud necrosis disease / Peanut spotted wilt / Groundnut ring mosaic

Symptoms

- » First symptoms are visible 2-6 weeks after infection as ring spots on leaves.
- » The newly emerging leaves are small, rounded or pinched inwards and rugose with varying patterns of mottling and minute ring spots.
- » Necrotic spots and irregularly shaped lesions develop on leaves and petioles.
- » Stem also exhibits necrotic streaks.
- » Plant becomes stunted with short internodes and short auxillary shoots.
- » Leaflets show reduction in size, distortion of the lamina, mosaic mottling and general chlorosis.
- » In advanced conditions, the necrosis of buds occurs.

Diseases of Groundnut   245

» Top bud is killed and necrosis spreads downwards.

» Drastic reduction in flowering and seeds produced are abnormally small and wrinkled with the dark black lesions on the testa.

Virus: It is caused by *Groundnut bud necrosis virus* (GBNV). The virus particles are spherical, 30 nm in diameter, enveloped, ssRNA with multipartite genome. It is also caused by *Tomato spotted wilt virus* (TSWV).

Disease cycle

The virus perpetuates in the weed hosts *viz., Bidens pilosa, Erigon bonariensis, Tagetes minuta* and *Trifolium subterraneum.* The virus is transmitted by thrips *viz., Thrips palmi, T. tabaci* and *Frankliniella* sp.

Management

» Adopt plant spacing of 15×15 cm.

» Remove and destory infected plants up to 6 weeks after sowing.

» Grow resistant varieties like ICGV 86030, ICGV 86031, ICGV 86032, ICGV 86033 and ICGV 86538.

» Application of monocrotophos 500 ml/ ha, 30 days after sowing either alone or in combination with AVP 10% (Anti Viral Principle) extracted from sorghum or coconut leaves.

» Spray the crop with AVP 10% at 500 lit / ha, ten and twenty days after sowing.

13. Indian peanut clump disease: *Peanut clump virus*

Earlier this disease was confused with groundnut rossette. Now it is recognized as a distinct virus causing clump disease. The leaves turn very dark and plants become severely stunted. The disease is soil borne and transmitted by a fungus, *Polymyxa graminis.* The pH of the soil affects transmission. It is also transmitted by seed. The virus is rod shaped, 190-245nm long × 21nm wide, not enveloped, ssRNA genome.

Other virus diseases of minor importance occurring on groundnut are:

» Peanut chlorotic streak (caused by *Caulimovirus*, occurs only in India),

» Peanut green mosaic and mottle (caused by a *Potyvirus*),

» Peanut stunt (caused by *Cucumovirus*),

» Groundnut chlorotic spot (caused by a *Potexvirus*),

» Groundnut eye spot (caused by *Potyvirus*) and groundnut ringspot.

Chapter - 27

Diseases of Sesame / Gingelly / Till
- *Sesamum indicum* L.

S. No.	Disease	Pathogen
1.	Root rot	*Macrophomina phaseolina*
2.	*Alternaria* leaf spot	*Alternaria sesami*
3.	Leaf spot	*Cercospora sesami*
4.	Wilt	*Fusarium oxysporum* f. sp. *sesami*
5.	Leaf and stem blight	*Phytophthora parasitica* var. *sesami*
6.	Powdery mildew	*Erysiphe cichoracearum*
7.	Bacterial blight	*Xanthomonas campestris* pv. *sesami*
8.	Bacterial leaf spot	*Pseudomonas sesami*
9.	Phyllody	*Candidatus* Phytoplasma asteris

1. Root rot / Stem rot / Charcoal rot

Symptoms

» The disease symptom starts as yellowing of lower leaves, followed by drooping and defoliation.

» The stem portion near the ground level shows dark brown lesions and bark at the collar region shows shredding.

» The sudden death of plants is seen in patches.

» In the grown-up plants, the stem portion near the soil level shows large number of black pycnidia.

» The stem portion can be easily pulled out leaving the rotten root portion in the soil.

» The infection when spreads to pods, they open prematurely and immature seeds shriveled and become black in colour.

» Minute pycnidia are also seen on the infected capsules and seeds.

» The rotten root as well as stem tissues contains a large number of minute black sclerotia.

» The sclerotia may also be present on the infected pods and seeds.

Fungus: *Macrophomina phaseolina* (Sclerotial stage: *Rhizoctonia bataticola*)

The pathogen produces dark brown, septate mycelium showing constrictions at the hyphal junctions. The sclerotia are minute, dark black and 110-130μm in diameter. The pycnidia are dark brown with a prominent ostiole. The conidia are hyaline, elliptical and single celled.

Disease cycle

The fungus remains dormant as sclerotia in soil as well as in infected plant debris in soil. The infected plant debris also carries pycnidia. The fungus primarily spreads through infected seeds which carry sclerotia and pycnidia. The fungus also spreads through soil-borne sclerotia. The secondary spread is through the conidia transmitted by wind and rain water.

Favourable conditions

» Day temperature of 30°C and above

» Prolonged drought followed by copious irrigation.

Management

» Grow resistant varieties like C 50, Gwalior 5, RT 1, RT 157 and Si 58.

» Seed treatment with carbendazim + thiram (1:1) at 2 g/ kg seed.

Diseases of Sesame 249

» Seed treatment with *Trichoderma viride* at 4 g/ kg.

» Apply farm yard manure or green leaf manure at 10 t/ ha or neem cake 150 kg/ ha.

» Spot drench with carbendazim at 1.0 g/ litre.

2. *Alternaria* leaf spot

Symptoms

» Initially small, circular, reddish brown spots (1-8mm) appear on leaves which enlarge later and cover large area with concentric rings.

» The lower surface of the spots are greyish brown in colour.

» In severe blighting defoliation occurs.

» Dark brown lesions can also be seen on petioles, stem and capsules.

» Infection of capsules results in premature splitting with shriveled seeds.

Fungus: *Alternaria sesami* Kawamura (Mohanty and Behera)

The mycelium of the fungus is dull brown and septate and produce large number of pale grey-yellow conidiophores which are straight or curved. The conidia are light olive coloured with transverse and longitudinal septa. There are around 3-5 septate and conidia are borne in chain over short conidiophore.

Disease cycle

The fungus is seed-borne and also soil-borne as it remains dormant in the infected plant debris.

Favourable conditions

Low temperature (20-25°C), high relative humidity, excessive rainfall and cloudy weather favour the disease.

Management

» Seed treatment with captan or thiram at 0.25% or carbendazim at 0.1%

» Spray twice with mancozeb at 0.25% or thiophanate methyl at 0.25% or carbendazim at 0.1%.

3. Leaf spot

Symptoms

- » The disease first appears on the leaves as minute water-soaked lesions, which enlarge to form round to irregular spots of 5-15 mm diameter on both the leaf surface.

- » The spots coalesce to form irregular patches of varying size leading to premature defoliation.

- » The infection is also seen on stem and petiole forming spots of varying lengths.

- » Dark linear spots also occur on pods causing drying shedding.

Fungus: *Cercospora sesame* Zimm.

The hypha of the fungus is irregularly septate, light brown and thick walled. Conidiophores are produced in cluster and are 1-3 septate, hyaline at the tip and light brown coloured at base. Conidia are elongated; 7-10 septate, hyaline to light yellow, broad at the base and tapering towards the apex.

Disease cycle

The fungus is externally and internally seed-borne. The fungus also survives in plant debris. Primary infection may be from the seeds and infected debris. The secondary spread is through wind-borne conidia.

Management

- » Seed treatment with carbendazin or thiram at 2 g/ kg.

- » Spray with mancozeb at 2 kg/ ha.

4. Wilt

Symptoms

- » The disease appears as yellowing, drooping and withering of leaves.

- » The plants gradually wither, show wilting symptoms leading to drying.

- » The infected portions of root and stem show long, dark black streaks of vascular necrosis.

Diseases of Sesame

Fungus: *Fusarium oxysporum* Schlecht. f. sp. *sesami*

The fungus produces macroconidia, microconidia and chlamydospores. Macroconidia are falcate shape, hyaline and 5-9 celled. Microconidia are hyaline, thin walled, unicellular and ovoid. The dark walled chlamydospores are also produced.

Disease cycle

The fungus survives in the soil in the infected plant debris. It is also seed-borne and primary infection occurs through infected seeds or through chlamydospores in soil. The secondary infection may be caused by conidia disseminated by rain splash and irrigation water.

Management

»   Seed treatment with thiram or carbendazim at 2 g/ kg.

»   Seed treatment with *Trichoderma viride* at 4 g/ kg.

»   Apply heavy doses of green leaf manure or farm yard manure.

5.   Leaf and stem blight

Symptoms

»   Black coloured lesions appear on the stem near the soil level.

»   The disease spreads further and affects branches and may girdle the stem, resulting in the death of the plant.

»   Leaves may also show water-soaked patches and spread till the leaves wither.

»   Infection may be seen on flowers and capsules.

»   Infected capsules are poorly developed with shriveled seeds.

Chromista: *Phytophthora parasitica* var. *sesame* Dastur

The pathogen produces non-septate, hyaline mycelium. The sporangiophores are hyaline and branched sympodially and bear sporangia. The sporangia are hyaline and spherical with a prominent apical papilla. The oospores are smooth, spherical and thick walled.

Disease cycle

The fungus can survive in the soil through dormant mycelium and oospores. The seeds also carry the fungus as dormant mycelium, which causes the primary infection. Secondary spread of the disease is through wind-borne sporangia.

Favourable conditions

» Prolonged rainfall, low temperature (25°C) and high relative humidity (above 90%).

Management

» Grow resistant varieties like ES 25, ES 105, ES 124, ES 156 and ES 190.

» Seed treatment with captan or thiram at 2 g/ kg or metalaxyl at 4 g/ kg.

» Avoid continuous cropping of sesamum in the same field.

» Remove and destrosy infected plant debris.

» Spray metalaxyl 1 kg/ ha.

6.   Powdery mildew

Symptoms

» Initially greyish-white powdery growth appears on the upper surface of leaves.

» When several spots coalesce, the entire leaf surface may be covered with powdery coating.

» In severe cases, the infection may be seen on the flowers and young capsules, leading to premature shedding.

» The severally affected leaves may be twisted and malformed.

» In the advanced stages of infection, the mycelial growth changes to dark or black because of development of cleistothecia.

Fungus: *Erysiphe cichoracearum* DC (Syn: *Oidium acanthospermi*)
Sphaerotheca fuliginea (Schltdl.) Pollacci; *Leveillula taurica* (Lév.) G. Arnaud

The fungi produce hyaline, septate mycelium which is ectophytic and sends haustoria into the host epidermis. Conidiophores arise from the primary mycelium and are short

Diseases of Sesame

and non septate bearing conidia in long chains. The conidia are ellipsoid or barrel-shaped, single celled and hyaline. The cleistothecia are dark, globose with the hyaline or pale brown myceloid appendages. The asci are ovate and each ascus produces 2-3 ascospores, which are thin walled, elliptical and pale brown in colour.

Disease cycle

The pathogen is an obligate parasite and disease perennates through cleistothecia in the infected plant debris in soil. The ascospores from the cleistothecia cause primary infection. The secondary spread is through wind-borne conidia.

Favourable conditions

- » Dry humid weather and low relative humidity.

Management

- » Grow resistant varieties like Si 44, Si 68, Si 71, Si 250, Si 1157, Si 1167, EC 20804, EC 20810 and EC 20835.

- » Remove the infected plant debris and destroy.

- » Spray wettable sulphur at 2.5 kg/ ha or triadimefon 1 ml/ lit repeat after 15 days.

7.   Bacterial blight

Symptoms

- » Initially water-soaked spots appear on the under surface of the leaf and then on the upper surface.

- » They increase in size, become angular and restricted by veins and dark brown in color.

- » Several spots coalesce together forming irregular brown patches and cause drying of leaves.

- » The reddish brown lesions may also occur on petioles and stem.

Bacterium: *Xanthomonas campestris* Dowson pv. *sesami*

The bacterium is a Gram negative rod with a monotrichous flagellum.

Disease cycle

The bacterium survives in the infected plant debris and in seeds. The secondary spread is by rain water.

Management

- » Grow resistant varieties like T 58.
- » Remove and burn infected plant debris.
- » Spray Streptomycin sulphate or oxytetracycline hydrochloride or streptocyclin at 100 g/ ha.

8. Bacterial leaf spot

Symptoms

- » The disease appears as water-soaked yellow specks on the upper surface of the leaves.
- » They enlarge and become angular as resticted by veins and veinlets.
- » The colour of spot may be dark brown with shiny oozes of bacterial masses.

Bacterium: *Pseudomonas sesami* Madhaiyan et al. 2017

The bacterium is gram negative aerobic rod with one or more polar flagella.

Disease cycle

The bacterium remains viable in the infected plant tissues. It is internally seedborne and secondary spread through rain splash and storms.

Management

- » Keep the field free of infected plant debris.
- » Spray with streptomycin sulphate or oxytetracycline hydrochloride or streptocyclin at 100 g/ ha.
- » Spray with copper based fungicides.

Diseases of Sesame

9. Phyllody

Symptoms

- » The symptom starts with vein clearing of leaves.
- » The disease manifests itself mostly during flowering stage, when the floral parts are transformed into green leafy structures, which grow profusely.
- » The flower is rendered sterile.
- » The veins of phylloid structure are thick and prominent.
- » The plant is stunted with reduced internodes and abnormal branching.

Bacterium: *Candidatus* Phytoplasma asteris

It is caused by pleomorphic mycoplasma like bodies present in sieve tube of affected plants, now designated as a phytoplasmal disease.

Disease cycle

The pathogen has a wide host range and survives on alternate hosts like *Brassica campestris* var. *toria, B. rapa, Cicer arietinum, Crotalaria* sp., *Trifolium* sp., *Arachis hypogaea* which serve as source of inoculum. The disease is transmitted by jassid, *Orosius albicinctus.* Optimum acquisition period of vector is 3-4 days and inoculation feeding period is 30 minutes. The incubation period of the pathogen in leaf hoppers may be 15-63 days and 13-61 days in sesame. Nymphs are incapable of transmitting the phytoplasma. Vector population is more during summer and less during winter months.

Management

- » Remove all the reservoir and weed hosts.
- » Avoid growing sesamum near cotton, groundnut and grain legumes.
- » Rogue out the infected plants periodically.
- » Spray monocrotophos or dimethoate at 500 ml/ ha to control the jassids
- » Soil treatment with thimet 10 G at 10 kg/ ha or phorate 10 G at 11 kg/ ha at the time of sowing.
- » Seed treatment with imidacloprid or carbosulfan protects the crop from all sucking pests including leaf hoppers for about a month
- » Spray dimethoate 30 EC at 500 ml/ ha at 30, 40 and 60 days after sowing to control the leaf hopper.

Chapter - 28

Diseases of Sunflower – *Helianthus annuus* L.

S. No.	Disease	Pathogen
1.	Root rot	*Rhizoctonia bataticola*
2.	Leaf blight	*Alternaria helianthi*
3.	Rust	*Puccinia helianthi*
4.	Head rot	*Rhizopus* sp.
5.	Powdery mildew	*Erysiphe cichoracearum*
6.	Basal rot	*Sclerotium rolfsii*
7.	Necrosis	*Tobacco streak virus*

1. Root rot or charcoal rot

Symptoms

» The pathogen is seed-borne and primarily causes seedling blight and collar rot in the initial stages.

» The grown up plants also show symptoms after flowering stage.

» The infected plants show drooping of leaves and death occurs in patches.

» The bark of the lower stem and roots shreds and are associated with a large number of sclerotia.

Diseases of Sunflower

» Dark coloured, minute pycnidia also develop on the lower portion of the stem.

Fungus: *Rhizoctonia bataticola* (Taub.) Butler;
Macrophomina phaseolina (Tassi) Goid (Pycnidial stage)

The fungus produces a large number of black, round to irregular shaped sclerotia. The pycnidia are dark brown to black with an ostiole and contain numerous single celled, thin walled, hyaline and elliptical pycnidiospores.

Disease cycle

The pathogen survives in soil and in infected crop residues through sclerotia and pycnidia. The pathogen is seed-borne and it serves as primary source of infection. Wind-borne conidia cause secondary spread. The soil borne sclerotia also spreads through rain splash, irrigation water and implements.

Favourable conditions

» Moisture stress and higher temperature favour development of the disease.

Management

» Closer planting of the seedling should be avoided.

» Optimum nutrition should be provided to maintain the plant vigour.

» Whenever the soil becomes dry and the soil temperature rises then irrigation should be provided.

» Seed treatment with *Trichoderma viride* formulation at 4 g/ kg seed.

» Soil application of *Trichoderma asperellum* / *T. harzianum* at 2.5 kg/ ha, preferably with organic amendments such as castor cake or neem cake or mustard cake at 500 kg/ ha.

» In endemic areas long crop rotation should be followed.

» Seed treatment with carbendazim or thiram at 2/ kg

» Spot drench with carbendazim at 500 mg/ litre.

258   Diseases of Field Crops and their Management

2. Leaf blight

Symptoms

- » The pathogen produces brown spots on the leaves, but the spots can also be seen on the stem, sepals and petals.

- » The lesions on the leaves are dark brown with pale margin surrounded by a yellow halo.

- » The spots later enlarge in size with concentric rings and become irregular in shape.

- » Several spots coalesce to show bigger irregular lesions leading to drying and defoliation.

Fungus: *Alternaria helianthi* (Hansf) Tubaki and Nishih.

The pathogen produces cylindrical conidiophores, which are pale grey-yellow coloured, straight or curved, geniculate, simple or branched, septate and bear single conidium. Conidia are cylindrical to long ellipsoid, straight or slightly curved, pale grey-yellow to pale brown, 1 to 2 septate with longitudinal septa.

Disease cycle

The fungus survives in the infected host tissues and weed hosts. The fungus is also seed-borne. The secondary spread is mainly through wind blown conidia.

Favourable conditions

- » Rainy weather.

- » Cool winter climate.

- » Late sown crops are highly susceptible.

Management

- » Deep summer ploughing.

- » Proper spacing

- » Clean cultivation and field sanitation.

Diseases of Sunflower

- » Grow resistant variety like EC 126184, EC 132846, ECC132847 and BSH-1.

- » Application of well rotten manures.

- » Practicing crop rotation.

- » Seed treatment with thiram or carbendazim at 2 g/ kg.

- » Spray mancozeb at 2 kg/ ha.

3. Rust

Symptoms

- » Small, reddish brown pustules (uredia) covered with rusty dust appear on the lower surface of bottom leaves.

- » Infection later spreads to other leaves and even to the green parts of the head.

- » In severe infection, when numerous pustules appear on leaves, they become yellow and dry.

- » The black coloured telia are also seen among uredia on the lower surface.

- » The disease is autoecious rust.

- » The pycnial and aecial stages occur on volunteer crops grown during off-season.

Fungus: *Puccinia helianthi* Schwein.

The uredospores are round or elliptical, dark cinnamon-brown in colour and minutely echinulated with 2 equatorial germpores. Teliospores are elliptical or oblong, two celled, smooth walled and cheshnut brown in colour with a long, colourless pedicel.

Disease cycle

The pathogen survives in the volunteer sunflower plants and in infected plant debris in the soil as teliospores. The disease spreads by wind-borne uredospores from infected crop.

Favorable conditions

- » Day temperature of 25.5°C to 30.5°C with relative humidity of 86 to 92% enhances intensity of rust attack.

260 Diseases of Field Crops and their Management

Management

- » Grow resistant varieties like EC 32361, EC 82819, MSFH 3, MSFH 9, MSFH 12 and BSH 1.

- » Crop rotation should be followed.

- » Previous crop remains should be destroyed.

- » Removal of crop residues.

- » Spray mancozeb at 2 kg/ ha.

4. Head rot

Symptoms

- » The affected heads show water soaked lesions on the lower surface, which later turn brown.

- » The discoloration may extend to stalk from head.

- » The affected portions of the head become soft and pulpy and insects are also seen associated with the putrified tissues.

- » The larvae and insects which attack the head pave way for the entry of the fungus which attacks the inner part of the head and the developing seeds.

- » The seeds are converted into a black powdery mass.

- » The head finally withers and droops down with heavy fungal mycelial nets.

Fungi: *Rhizopus* sp. (*Rhizopus arrhizus* (Fisher), *Rhizopus stolonifer* (Ehrenb.) Vuill., *Rhizopus nigricans* Ehrenb. and *Rhizopus microspores* Tiegh.)

Pathogen produces dark brown or black coloured, non-septate hyphae. It produces many aerial stolens and rhizoids. Sproangia are globose and black in colour with a central columella. The sporangiospores are aplanate, dark coloured and ovoid.

Favourable conditions

- » Prolonged rainy weather at flowering.

- » Damages caused by insects and caterpillars.

Diseases of Sunflower 261

Disease cycle

The fungus survives as a saprophyte in host debris and other crop residues. The disease is spread by wind blown spores.

Management

- » Seed treatment with thiram or carbendazim at 2 g/ kg.
- » Control the caterpillars feeding on the heads.
- » Spray the head with mancozeb at 2 kg/ ha during intermittent rainy season and repeat after 10 days, if the humid weather persists.

5. Powdery mildew

Symptoms

- » White to grey mildew on the upper surface of older leaves.
- » As plant matures black pin head sized are visible in white mildew areas.
- » The affected leaves more luster, curl, become chlorotic and die.

Fungus: *Erysiphe cichoracearum* DC.

Favorable conditions

- » The disease is more under dry condition to the end of the winter months.

Management

- » Complete field and crop sanitation.
- » Early varieties should be preferred.
- » Removal of infected plant debris.
- » Spray calixin 1 lit/ ha or wettable sulphur 2 kg/ ha.

6. Basal rot

Symptoms

- » Initial symptoms of the disease appear 40 days sowing.
- » The infected plants can be identified by their sickly appearance.
- » Plants dry up due to the disease infestation.
- » The lower portion of stem is covered with white or brownish white fungal colonies.
- » In extreme cases the plants wilts and dies.
- » Dark brown lesions appear on the base of the stem near ground level, leading to withering.
- » Large numbers of sclerotia are seen.

Fungus: *Sclerotium rolfsii* Sacc.

Favourable conditions

- » Infection occurs in the crop in the month of July and August.
- » The fungus survives through sclerotina in soil and plant debris.

Management

- » Deep summer ploughing.
- » Complete field and crop sanitation.
- » Use of resistant or tolerant varieties.
- » Apply *Trichoderma* on seed and soil to reduce wilt.
- » Apply *Coniothyrium minitans* bioagent in soil before sowing.
- » Seed treatment with *Pseudomonas fluorescens* or *Pseudomonas putida* strains protect sunflower from *Sclerotinia* infection during seedling stage.
- » Seed treatment with captan or thiram at the rate of 3 g/ kg.
- » Drenching the base of the plant with chestnut compound 3 g per litre of water.
- » Seed treatment with carbendazim at 0.2% followed by the addition of *Trichoderma*

Diseases of Sunflower

harzianum 10 g/ kg soil and spraying carbendazim at 0.2 % to 15 days old seedling.

7.   Necrosis

Symptoms

> » Characterised by the sudden necrosis of part of lamina followed by twisting of leaves and systemic mosaic.

> » Necrosis of lamina of the lamina, petiole, stem floral calyx and corolla.

Virus: *Tobacco streak virus* an Ilarvirus 25-28 nm, tripartite genome encapzidated separately

Disease cycle

Virus spreads through transmission by thrips *Frankliniella schultzii*. Weed hosts serve as natural virus reservoirs. Long and continuous dry spell increases the disease incidence.

Management

> » Removal of weed hosts

> » Management of veetor population

> » Changing planting dates

Minor diseases

> » Downy mildew   - *Plasmopara halstedii*

> » *Cercospora* leaf spot   - *Cercospora helianthicola*

> » *Septoria* leaf spot   - *Septoria helianthi*

> » *Verticillium* wilt   - *Verticillium dahliae*

> » *Sclerotinia* wilt/ stem rot - *Sclerotinia sclerotiorum*

Chapter - 29

Diseases of Safflower
- *Carthamus tinctorius* L.

S. No.	Disease	Pathogen
1.	Leaf blight	*Alternaria carthami*
2.	Wilt	*Fusarium oxysporum* f. sp. *carthami*
3.	Rust	*Puccinia carthami*
4.	Mosaic	*Cucumber mosaic virus*

1.   Leaf blight

It is the most destructive disease and appears in a severe form wherever safflower is grown.

Symptoms

- » Dark brown lesions measuring 2-5mm in diameter are first found on hypocotyls and cotyledons.

- » The disease is severe on leaves and occasionally attacks stem and flowers.

- » Minute brown to dark brown spots with concentric rings of 1-2mm appear on leaves.

- » The centre of the spot is light brown with a dark brown margin.

- » Elongated black lesions can be seen on the petiole and stem.

Diseases of Safflower 265

» The fungal infection on flower buds leads to drying and shedding.

» Seeds also may be affected.

» Dark sunken lesions are produced on the testa.

Fungi: *Alternaria carthami* S. Chowdhury and *Alternaria alternata* (Fr.) Keissl.

The mycelium of the fungus is sub-hyaline initially and become brown coloured on maturity. The conidiophores are stout, erect, rigid, unbranched, septate and arise singly or in clusters. The conidia are 3-11 celled with irregular shape, light brown in colour with a long beak.

Disease cycle

The fungus is externally seed-borne and also survives in plant debris. The disease spread is through windblown conidia.

Management

» Collect and destroy infected plant debris.

» Grow resistant varieties like NS 133, NS 1015, NS 1016, NS 1021 and EC 32102.

» Seed treatment with thiram or captan at 3 g/ kg or carbendazim at 0.1%.

» Hot water treatment of seed at 50°C for 30 minutes.

» Spray mancozeb or zineb at 0.2% or carbendazim at 0.1%.

2. Wilt

Symptoms

» In seedling stage cotyledonary leaves show small brown spots either scattered or arranged in a ring on the inner surface and they may be shrivelled or rolled or curved.

» Symptoms become apparent when plants are in 6-10 leaf stage as yellowing of leaves followed by wilting,

» Epinasty and vascular browning develop in acropetal succession.

» In older plants the lateral branches on one side may be killed while the remainder of the plant remains free from the disease.

266 Diseases of Field Crops and their Management

» Infected plants produce small sized flower heads which are partially blossomed.

» Most of the ovaries fail to develop seeds or they may form blackish, small, distorted chaffy seeds.

Fungus: *Fusarium oxysporum* f. sp. *carthami* Klis. & Houston, (1963)

The fungus produces hyaline, septate mycelium. Microconidia are hyaline, small, elliptical or curved, single celled or two celled. Macroconidia are also hyaline, thin walled, linear, curved or fusoid, pointed at both ends with 3-4 septa. The fungus also produces thick walled, spherical or oval, terminal or intercalary.

Disease cycle

» The fungus survives in seed, soil and infected plant debris. The primary spread is by soilborne chlamydospores and also by seed contaminant.

» The secondary spread in the field is through irrigation water and implements.

Favourable conditions

» Acid soil, high nitrogen, warm moist weather and temperature of 15-20°C.

Management

» Avoid growing safflower in low lying areas

» Collection and destruction of plant debris.

» Follow crop rotation with sorghum

» Grow wilt resistant / tolerant hybrids DSH 129, NARI-NH-1 and varieties A1, PBNS 40 and NARI 6 in endemic areas.

» Seed treatment with thiram or captan at 3 g/ kg or carbendazim at 0.1% or *T. viride* at 10 g/ kg.

3. Rust

Symptoms

» The fungus attacks cotyledons, young leaves, tender stems and underground parts.

Diseases of Safflower

» Infection of the cotyledons is seen as yellow discoloration accompanied by drooping and wilting.

» The pustules (uredosori) are chestnut brown in colour, erumpent and scattered throughout the leaves.

» Later in the season, black teliosori are formed on the same spots.

» Seedlings sometimes die suddenly without exhibiting symptoms in the aerial parts.

» Infection on the hypocotyl causes hypertrophy of the tissues due to accumulation of mycelium between cells.

» Stem girdling occurs in older plants.

» The rust pustules also appear on tap root and lateral roots.

Fungus: *Puccinia carthami Corda* (*Puccinia calcitrapae* var. *centaureae*) or *P. verruca* or *Aecidium carthami*

The fungus is an obligate parasite with autoecious life cycle in safflower. Uredia and telia are produced and pycnial and aerial stages are unknown. Uredospores are single celled, light brown coloured and echinulate. Teliospores are globose to broadly ellipsoid, two celled, chestnut brown in colour, thick walled with hyaline pedicels.

Disease cycle

The fungus remains on the seeds and infected crop debris in the soil as teliospores for more than a year. The fungus also produces uredial and telial stages in the collateral host *Carthamus oxyacantha* and this also serves as primary source of infection in addition to dormant teliospores in soil. The secondary spread occurs through wind-borne uredospores.

Management

» Grow resistant varieties like Sagaramuthyalu, Manjeera APPR 1, APPR 3, PCM 1, PCM 2 and PCN 1.

» Seed treatment with thiram or captan at 3 g/ kg or carbendazim at 2 g/ kg.

» Remove and destroy the plant debris in the soil.

» Rogue out the collateral host.

> » Spray wettable sulphur or mancozeb at 0.2%.

4. Mosaic

Symptoms

> » In CMV infected safflower plants, young leaves show irregular yellow or light green patches alternating with normal green areas.

> » Leaves may become blistered and distorted and infected plants are stunted.

> » In few plants primary leaves are produced, forming a rosette of leaves exhibiting mosaic mottling and from the centre of this the axis bearing secondary leaves is produced.

Virus: *Cucumber mosaic virus* (CMV)

Disease cycle

The virus can infect a number of wild and cultivated plants and is transmitted by aphid, *Myzus persicae*.

Management

> » Rogue out and destroy infected plants

> » Spray insecticides like monocrotophos 1.5 ml or dimethoate 2 ml for the control of aphid vectors

Chapter - 30

Diseases of Castor -*Ricinus communis* L.

S. No.	Disease	Pathogen
1.	Seedling blight	*Phytophthora parasitica*
2.	Rust	*Melampsora ricini*
3.	Leaf blight	*Alternaria ricini*
4.	Brown leaf spot	*Cercospora ricinella*
5.	Powdery mildew	*Leveillula taurica*
6.	Stem rot	*Macrophomina phaseolina*
7.	Wilt	*Fusarium oxysporum*
8.	Bacterial leaf spot	*Xanthomonas campestris* pv. *ricinicola*

1.   Seedling blight

Symptoms

- » The disease appears circular, dull green patch on both the surface of the cotyledon leaves.

- » It later spreads and causes rotting.

- » The infection moves to stem and causes withering and death of seedling.

270   Diseases of Field Crops and their Management

» In mature plants, the infection initially appears on the young leaves and spreads to petiole and stem causing black discoloration and severe defoliation.

Chromista: *Phytophthora parasitica* Dastur

The pathogen produces non-septate and hyaline mycelium. Sporangiophores emerge through the stomata on the lower surface singly or in groups. They are unbranched and bear single celled, hyaline, round or oval sporangia at the tip singly. The sporangia germinate to produce abundant zoospores. The fungus also produces oospores and chlamydospores in adverse seasons.

Disease cycle

The pathogen remains in the soil as chlamydospores and oospores which act as primary source of infection. The fungus also survives on other hosts like potato, tomato, brinjal, sesamum etc. The secondary spread takes place through wind borne sporangia.

Favourable conditions

» Continuous rainy weather.

» Low temperature (20-25°C).

» Low lying and ill drained soils.

Management

» Remove and destroy infected plant residues.

» Avoid low-lying and ill drained fields for sowing.

» Seed treatment with thiram or captan at 4 g/ kg.

» Spray mancozeb 0.25% or copper oxychloride 0.5% or carbendazim 0.1% or metalaxyl 0.4%.

2.   Rust

Symptoms

» Minute, orange-yellow coloured, raised pustules appear with powdery masses on the lower surface of the leaves and the corresponding areas on the upper surface of

Diseases of Castor   271

the leaves are yellow.

» Often the pustules are grouped in concentric rings and coalesce together to for drying of leaves.

Fungus: *Melampsora ricini* Pass. ex E.A. Noronha

The pathogen produces only uredosori in castor plants and other stages of the life cycle are unknown. Uredospores are two kinds, one is thick walled and other is thin walled. They are elliptical to round, orange-yellow coloured and finely warty.

Disease cycle

The fungus survives in the self sown castor crops in the off season. It can also survive on other species of *Ricinus*. The fungus also attacks *Euphorbia obtusifolia, E.geniculata* and *E.marginata*. The infection spreads through airborne uredospores.

Management

» Rogue out the self-sown castor crops and other weed hosts.

» Spray mancozeb at 2 kg/ ha or propiconazole 1 lit/ ha.

3.   Leaf blight

Symptoms

» All the aerial parts of plants *viz.*, leaves, stem, inflorescences and capsules are liable to be attacked by the pathogen.

» Irregular brown spots with concentric rings form initially on the leaves and covered with fungal growth.

» All the spots coaleasce to form big patches, premature defoliation occurs.

» The stems, inflorescences and capsules are also show dark brown lesions with concentric rings.

» On the capsules, initially brown sunken spots appear, enlarge rapidly and cover the whole pod.

» The capsules crack and seeds are also got infected.

Fungus: *Alternaria ricini* (Yoshii) Hansf.

The pathogen produces erect or slightly curved, light grey to brown conidiophores, which are occasionally in groups. Conidia are produced in long chains. Conidia are obclavate, light olive in colour with 5-16 cells having transverse and longitudinal septa with a beak at the tip.

Disease cycle

The pathogen survives on hosts like *Jatropha pandurifolia* and *Bridelia hamiltoniana.* The pathogen is externally and internally seed-borne and causes primary infection. The secondary infection is through air-borne conidia.

Favourable conditions

> » High atmospheric humidity 85-90% and low temperature of 16-20°C.

Management

> » Seed treatment with captan or thiram at 2 g/ kg.

> » Remove the reservoir hosts periodically.

> » Spray mancozeb 0.25% or copper oxychloride 0.5% or carbendazim 0.1% or metalaxyl 0.4%.

4.   Brown leaf spot

Symptoms

> » The disease appears as minute brown specks surrounded by a pale green halo.

> » The spots enlarge to greyish white centre portion with deep brown margin.

> » The spots may be 2-4 mm in diameter and when several spots coalesce, large brown patches appear but restricted by veins.

> » Infected tissues often drop off leaving shot-hole symptoms.

> » In severe infections, the older leaves may be blighted and withered.

Fungus: *Cercospora ricinella* Sacc. & Berl.

Diseases of Castor

The pathogen hyphae collect beneath the epidermis and form a hymenial layer. Clusters of conidiophores emerge through stomata or epidermis. They are septate and un branched with deep brown base and light brown tip. The conidia are elongated, colourless, straight or slightly curved, truncate at the base and narrow at the tip with 2-7 septa.

Disease cycle

The pathogen remains as dormant mycelium in the plant debris. The disease mainly spreads through wind borne conidia.

Management

- » Spray Bordeaux mixture 1% or copper oxy chloride 0.2% may help to bring the disease under check; but where the cultures of Eri-silk worm are maintained on castor plants, spraying would not be desirable.
- » Spray with mancozeb 2 g/ lit or carbendazim 500 g/ ha at 10-15 day interval.
- » Seed treatment with thiram or captan 2 gm/ kg.

5. Powdery mildew

Symptoms

- » It is characterized by typical mildew growth which is generally confined to the under-surface of the leaf.
- » When the infection is severe the upper-surface is also covered by the whitish growth of the fungus.
- » Light green patches, corresponding to the diseased areas on the under surface, are visible on the upper side especially when the leaves are held against light.

Fungus: *Leveillula taurica* Lev.

Management

- » Grow resistant variety like Jwala.
- » Spray wettable sulphur 2 g/ lit at 15 days interval, from 3 months after sowing.
- » Spray wettable sulphur 3 g/ lit or hexaconazole 1 ml/ litre at fortnight intervals.

274 Diseases of Field Crops and their Management

6. Stem rot / Root rot

Symptoms

» Small brown depressed lesions on and around nodes.

» Increase in size on both directions causing 2-20 cm necrotic area.

» Lesions often coalesce and girdle the stem causing leaf drop.

» Drying and death starts from apex and progress.

» Infected capsules discoloured and drop easily.

» Sudden wilting of plants in patches under high moisture stress coupled with high soil temperature.

» Plant exhibit symptoms of drought and drooping of leaves.

» At ground level black lesions are formed on the stem.

» Young leaves curl inwards with black margins and drop off later, such branches Die-back.

» Entire branch and top of the plant withers.

Fungus: *Macrophomina phaseolina*

Management

» Grow tolerant and resistant varieties like Jyothi, Jwala, GCH-4, DCH-30 and SHB-145.

» Avoid water logging.

» Destruction of crop debris.

» Selection of healthy seed.

» Providing irrigation at critical stages of the crop.

» Treat the seed with thiram at 2g/kg or carbendazim at 2g/ kg.

» Seed treatment with *Trichoderma viride* formulation at 4g/ kg.

» Furrow application of *Trichoderma harzianum* 5 kg/ 500 kg of mustard cake.

» Soil drenching with Carbendazim (1g/1 litre of water) 2-3 times at 15 days interval.

Diseases of Castor

7. Wilt

Symptoms

- » When seedlings are attacked cotyledonary leaves turn to dull green colour, wither and die subsequently.
- » Leaves are droop and drop off leaving behind only top leaves.
- » Diseased plants are sickly in appearance.
- » Wilting of plants, root degeneration, collar rot, drooping of leaves and necrosis of affected tissue and finally leading to death of plants.
- » Necrosis of leaves starts from margins spreading to interveinal areas and finally to the whole leaf.
- » Spilt open stem shows brownish discolouration and white cottony growth of mycelia much prominently in the pith of the stem.

Fungus: *Fusarium oxysporum* Schlecht. emend. Snyder & Hansen

Management

- » Selection of disease free seeds.
- » Grow tolerant and resistant varieties like Jyothi, Jwala, GCH-4 DCH-30 and SHB 145.
- » Avoid water logging and burning of crop debris after harvest.
- » Green manuring and intercropping with red gram
- » Seed treatment with thiram at 2g/ kg or carbendiazim at 2g/ kg seed.
- » Seed treatment with 4g of *Trichoderma viride* talc formulation.
- » Multiplication of 2kg of *T.viride* formulation by mixing in 50kg farm yard manure
- » Sprinkling water and covering with polythene sheet for 15days and then applying between rows of the crops is helpful in reducing the incidence.

8. Bacterial leaf spot

Symptoms

- » The pathogen attacks cotyledons, leaves and veins and produces few to numerous small round, water-soaked spots which later become angular and dark brown to jet black in color.

- » The spots are generally aggregated towards the tip.

- » At a later stage the spots become irregular in shape particularly when they coalesce and areas around such spots turn pale-brown and brittle.

- » Bacterial ooze is observed on both the sides of the leaf which is in the form of small shining beads or fine scales.

Bacterium: *Xanthomonas campestris* pv. *ricini* (Yoshii & Takimoto) Vauterin

Management

- » Field sanitation help in minimizing the yield loss as pathogen survives on seed and plant debris.

- » Hot water treatment of seed at 58°C to 60°C for ten minutes.

- » Grow tolerant varieties.

- » Spray Copper oxychloride 2kg/ ha or Streptocycline 100g/ ha or Paushamycin 250g/ ha.

Chapter - 31

Diseases of Linseed / Flax
- *Linum usitatissimum* L.

S. No.	Disease	Pathogen
1.	Rust	*Melampsora lini*
2.	Wilt	*Fusarium oxysporum* f. sp. *lini*
3.	Pasmo (Stem banding) disease	*Septoria linicola*
4.	Powdery mildew	*Oidium lini*
5.	Stem break and browning	*Aureobasidium pullulans* var. *lini*
6.	Seedling blight and root rot	*Rhizoctonia solani*
7.	*Sclerotinia* stem rot	*Sclerotinia sclerotiorum*
8.	Aster Yellows	Aster Yellows Phytoplasma
9.	Crinkle	*Oat blue dwarf virus*

1.   Rust

Rust is potentially the most destructive disease affecting flax. The last major rust epidemic occurred in the 1970's. Although it is effectively controlled by genetic resistance, it remains a potential threat to flax production as it can survive locally and complete its life cycle on flax, thus having the ability to produce new races that attack hitherto resistant varieties.

Symptoms

» Rust is readily recognized by the presence of bright orange and powdery pustules, also called uredia.

» Rust pustules develop mostly on leaves, but also on stems and bolls.

» The pustules produce numerous urediospores which are airborne and cause new cycles of infections during the season.

» Spread and infections are favored by high humidity during cool nights, warmer day temperatures and on plants growing vigorously.

» As the season progresses, the orange pustules turn black and produce overwintering telia and teliospores.

» The black pustules are most common on stems.

Fungus: *Melampsora lini* (Ehrenb.) Lev., (1847)

The causal organism is *Melampsora lini*, a fungus that overwinters by means of teliospores on flax debris. Early infections produce the aecial stage with aeciospores on volunteer flax seedlings which subsequently produce the uredial stage. Urediospores can cycle through several generations during the growing season resulting in completely defoliated flax plants and reduction of seed yield and fiber quality. Flax rust completes its life cycle on the flax plant, unlike many other rusts that require an alternate host.

Management

» Complete control is achieved by the use of rust-resistant varieties.

» Grow disease resistant varieties like Neelum, T 1, Hira, T 126, K2, LC 36, NP (RR) 9, 10, 56, 95, 218, 279B 279K3, 368, 381, 389, 415 and 501 is the only effective method to control the disease.

» Dust with sulphur 20-25 kg/ ha.

» Spray mancozeb 2.5 g/ lit or sulphur 2-3 g/ lit or carbendazim 2g/ lit or hexaconazole 1.5 ml/ lit.

Diseases of Linseed

2. Wilt

Flax wilt or fusarium wilt is caused by the seedborne and soilborne fungus *Fusarium oxysporum* f. sp. *lini*. The fungus invades plants through the roots at any growth stage during the growing season and continues infection inside the water-conducting tissue of the root. This interferes with water uptake and warm weather therefore aggravates plant symptoms from the disease.

Symptoms

> » Early infections may kill flax seedlings shortly after emergence, while delayed infections cause yellowing and wilting of leaves, followed by browning and death of plants.

> » Roots of dead plants turn ashy grey.

> » The tops of wilted plants often turn downward and form a "shepherd's crook".

> » Affected plants occur more commonly in patches but may also be scattered throughout the field.

> » The fungus persists in the soil, as mycelia and spores survive for many years in debris of flax and other organic matter in the soil.

> » Wind-blown and water run-off soil may spread the fungus from one field to another.

Fungus: *Fusarium oxysporum* f. sp. *lini* (Bolley) W.C. Snyder & H.N. Hansen

Management

> » Grow resistant varieties.

> » Crop rotation of at least three years.

> » Seed treatment with carbendazim 0.2% or mancozeb 0.2% or captan 0.2% or copper hydroxide 0.2%.

3. Pasmo (Stem banding) disease

Symptoms

> » Pasmo is characterized by circular brown lesions on the leaves and brown to black infected bands that alternate with green and healthy bands on the stem.

» Infected flax tissue is characterized by tiny black pycnidia which are the fruiting bodies of the fungus.

» The debris carries numerous pycnidia which overwinter and produce masses of spores that cause the initial infections on leaves and stems.

» Spores are dispersed by rain and wind.

» High moisture and warm temperatures favor the disease.

» Lodging favors the development of pasmo, because of increased humidity within the crop canopy and this may result in patches of dead plants completely covered with the fungus.

Fungus: *Septoria linicola* (Speg.) Garass., (1938)

The causal organism of this disease is *Septoria linicola*, a fungus that attacks above-ground parts of flax and overwinters in the soil on infected flax stubble. Flax is susceptible to pasmo from the seedling stage to maturity. Epidemics can occur early in the season when favorable conditions of high humidity with frequent rain showers prevail. Pasmo can cause defoliation; premature ripening and can weaken the infected pedicels resulting in heavy boll-drop under rain and wind conditions. Depending on the earliness and severity of the infection, pasmo reduces seed yield as well as seed and fiber quality. Commercial flax varieties lack resistance to this pathogen.

Management

» Spray carbendazim 0.1% or chlorothanil 0.2%.

4. Powdery mildew

This disease was first reported in Western Canada in 1997. Powdery mildew has spread quickly and its incidence and severity have increased sharply in Manitoba and Saskatchewan.

Symptoms

» The symptoms are characterized by a white powdery mass of mycelia that start as small spots and rapidly spread to cover the entire leaf surface.

» Heavily infected leaves dry up, wither and die.

» Early infections may cause complete defoliation of flax plants.

Diseases of Linseed 281

Fungus: *Oidium lini* Bondartsev

The causal agent is the fungus *Oidium lini* and little is known about the overwintering and host range of this fungus in Western Canada. Early infections may cause severe defoliation of the flax plant and reduce seed yield and quality. Some flax varieties are resistant/moderately resistant to this disease.

Management

» Spray recommended wettable sulphur around flowering time.

5.   Stem break and browning

Stem break and browning are phases of a disease caused by the seedborne and soilborne fungus, *Aureobasidium pullulans* var. *lini*, also called *Polyspora lini*. This disease is of minor importance in Western Canada; however, it may cause some damage in the Parkland regions of Saskatchewan and Alberta in some years.

Symptoms

» Stem break is the first conspicuous disease symptom.

» Development of a canker at the stem base weakens the plant.

» The stem may break at this point when the plants are still young, or at a later stage.

» Plants may remain alive after stem breakage, but any seed produced may still be lost in harvesting as seed produced will be smaller and thin.

» Initial infections in spring may start from spores produced on diseased stubble and are spread by wind and rain.

» Infections may start during seedling emergence when seed coats of diseased seed are lifted above the ground and the fungus produces the first cycle of spores of the season.

» The browning phase is initiated by infections on the upper part of the stem that appear as oval or elongated brown spots, often surrounded by narrow, purplish margins.

» The spots may coalesce, and leaves and stem turn brown.

» Patches of heavily infected plants appear brown, giving the disease the name of 'browning'.

282   Diseases of Field Crops and their Management

» The fungus may penetrate bolls as well as seeds, or may produce spores on the seed surface. Affected seeds may remain viable.

Fungus: *Aureobasidium pullulans* var. *lini* (Laff.) W.B. Cooke

Management

» Use of disease-free seed produced by healthy plants is the most important control measure.

» Spray copper oxychloride 0.25% or copper hydroxide 0.2%.

6. Seedling blight and root rot

Symptoms

» Blighted seedlings turn yellow, wilt and die.

» Infected seedlings may occur singly or in patches.

» Seedling blight may be inconspicuous and gaps in the row may be the principal sign of disease occurrence.

» Roots of recently affected plants show red to brown lesions, and may later turn dark and shrivel.

» Diseased plants are often difficult to distinguish from those killed by the wilt fungus.

» Root rot symptoms usually appear in plants after the flowering stage.

» Plants may wilt on warm days and turn brown prematurely.

» Plants with root rot usually set little or no seed.

Fungus: *Rhizoctonia solani* Kuhn.

Management

» Seedling blight and root rot can be controlled by a combination of farm practices.

» Use certified seed of a recommended variety.

» Reduce cracking of seed by adjusting combine settings during harvest.

» Treat the seed with a carbendazim 0.2%.

Diseases of Linseed 283

» Practice a crop rotation of at least three years between flax crops and plant in a field that is distant from fields sown to flax in the previous year.

7. *Sclerotinia* stem rot

Symptoms

» The symptoms are water-soaked longitudinal lesions on the stems girdling the stems resulted in bleaching, shredding of the stem, breakage and lodging in Sclerotinia heavily infested fields.

» Mycelia grow on the stem surface and cylindrical shaped sclerotia are formed inside the stem.

Fungus: *Sclerotinia sclerotiorum* (Lib.) de Bary

The causal agent is *Sclerotinia sclerotiorum* which is a widespread pathogen causing diseases on canola, sunflower, soybean, leguminous crops, and 100s of plant species. This fungus survives for 3-4 years in the soil as compact masses of mycelia called sclerotia. The infection in flax is caused by mycelial infection on plants touching the infested soil in heavily lodged flax. No evidence of airborne ascospores infection of this pathogen in flax.

Management

» Crop rotation

» Summer ploughing

» Apply neem cake 150 kg/ ha.

8. Aster Yellows

Symptoms

» Aster yellows symptoms include yellowing of the top part of the plant, conspicuous malformation of the flowers and stunted growth.

» All flower parts including the petals are converted into small, yellowish green leaves.

» Diseased flowers are sterile and produce no seed.

» The severity of the disease depends at the stage plants become infected and the number of insect vectors that carry the pathogen.

Diseases of Field Crops and their Management

» The mycoplasma-like organism overwinters in perennial broadleaved weeds and crops, but most infections are carried by leafhoppers.

» The six-spotted leafhopper is the main vector which transmits the phytoplasma organism that causes aster yellows in flax, canola, sunflower and some weeds.

Bacterium: Aster Yellows Phytoplasma

Management

» Seed early to avoid the migrating leafhoppers in mid to late season.

» Early summer migration of leafhoppers occurs when abnormal warm weather prevails early in the growing season thus resulting in major epidemic and yield loss.

9. Crinkle

Symptoms

» Crinkle is caused by a virus called oat blue dwarf that also causes disease in oats, wheat, and barley.

» The symptoms are characterized by a conspicuous puckering of leaves, stunted growth and reduced branching.

» Flowering may appear normal but seed production is reduced.

» Like aster yellows, crinkle is a disease of flax that depends for infection via transmission by the six-spotted leafhopper.

Virus: *Oat blue dwarf virus*

Management: Seed early to avoid migrating leafhoppers in mid to late season.

Minor Diseases

» **Seedling and stem blight**: *Alternaria linicola*

» **Anthracnose**: *Colletotrichum lini*

» **Root rot**: *Phoma exigua* and

» **Dieback**: *Selenophoma linicola*

» **Stem mold**: *Sclerotinia sclerotiorum*

Chapter - 32

Diseases of Mustard
- *Brassica juncea* (L.) Czern.

S. No.	Disease	Pathogen
1.	Damping-off	*Phytophthora cactorum*
2.	Wilt	*Fusarium oxysporum* f. sp. *conglutinans*
3.	*Alternaria* spot	*Alternaria brassicae*
4.	*Sclerotium* foot rot	*Sclerotium rolfsii*
5.	Downy mildew	*Hyaloperonospora parasitica* var. *lini*
6.	Powdery mildew	*Erysiphe cruciferarum*
7.	Black leg	*Leptosphaeria maculans*
8.	White blister	*Albugo candida*
9.	Club root	*Plasmodiophora brassicae*
10.	Bacterial black rot	*Xanthomonas campestris* pv. *campestris*
11.	Phyllody and Aster Yellows	*Candidatus* Phytoplasma asteris
12.	Mosaic	*Cauliflower mosaic virus*

1. Damping-off, Wirestem, Brown Girdling and Root Rot

Seedling diseases including damping-off, seedling blight, wirestem, and root rot affect *Brassica* and other plant species including pulses and flax.

Symptoms

- » A necrotic lesion 1–2 cm long may be seen at the base of the stem, with girdling sometimes taking place near the soil level.

- » The taproot may be discolored and sometimes wire-stem symptoms may be seen.

- » Salmon-colored spore masses of *Fusarium* are often observed on affected tissues.

- » Girdling of the main root may take place, which may lead to loss of the entire root system.

- » Damping-off and seedling blight are mostly encountered due to the use of infested seed.

- » Primary lesions consisting of small, circular necrotic spots, along with secondary lesions with large irregular borders, appear on leaves.

Damping off: *Phytophthora cactorum* (Leb. and Cohn) Schröeter and/or *Pythium* spp.

Wirestem and brown girdling root rot: Anamorph: *Rhizoctonia solani* Kuhn
Teleomorph: *Thanatephorus cucumeris* (A. B. Frank) Donk

Management

- » Seed treatment with thiophanate methyl 70 WP at 2 g/ kg.

- » Seed treatment with metalaxyl 35 SD 6 g/ kg or carbendazim 1 g/ kg.

2. Wilt

Symptoms

- » The leaves of the affected plants show drooping, vein clearing, and chlorosis, followed by wilting, drying, resulting in the death of the plant.

- » The symptoms progress from the base to upward.

- » The expression of the disease symptoms varies with the age of the plants.

- » In the early stage of development, affected plants do not show all the typical symptoms.

- » Roots of the diseased plants show no external abnormality or decay of the tissue until the plants are completely dried.

Diseases of Mustard

» Vascular tissues of stem and root show the presence of the mycelium and/or microconidia of the pathogen.

» Such tissues show browning of their walls and their plugging with a dark gummy substance, which is one of the characteristic symptoms of vascular wilts.

» At later stages of the disease, epidermis of roots sloughs off.

» The diseased plants eventually collapse and die.

Fungus: *Fusarium oxysporum* f. sp. *conglutinans* (Wr.) Snyder and Hansen.

Management

» Seed treatment with carbendazim at 0.1%.

» Extracts of plants *Vitex trifolia* and *Artemisia nilagirica* were used.

3.   *Alternaria* spots (Black spot, Gray Leaf Spot, Pod Spot)

Symptoms

» The disease attacks on the lower leaves as small circular brown necrotic spots which slowly increase in size.

» Many concentric spots coalesce to cover large patches showing blightening and defoliation in severe cases.

» Circular to linear, dark brown lesions also develop on stems and silique, which are elongated at later stage.

» Infected pods produce small, discolored and shriveled seeds.

» Spots produced by *Alternaria brassicae* appear to be usually gray in color compared with black sooty velvety spots produced by *Alternaria brassicicola*.

» Spots produced by *Alternaria raphani* show distinct yellow halos around them.

Fungi: *Alternaria brassicae* (Berk.) Sacc., *Alternaria brassicola* (Schwein.) Wiltshire, *Alternaria raphani* Groves & Skolko, *Alternaria longipes* and *Alternaria napiforme*.

Favourable conditions

» The conidia are dark, obclavate, muriform and are borne singely or in chains.

» In temperate countries these fungi are seed borne and cause shriveling of the seeds

» Spores and mycellium in diseased plant debria serve as the means of perination.

» Conidia are abundantly formed in moist atmosphere.

» Conidia are disseminated by wind.

» In tropical conditions the seed borne innoculum is devitalized soon after the harvest of the seeds in summer.

Management

» Grow resistant varieties like RLM 619.

» Spray with mancozeb 0.2% or captafol 0.2% or chlorothalonil 0.2% at the flowering time of the crop nearly 3-4 times at 10-15 days interval.

» Seed treatment with carbendazim and foliar spray of metalaxyl + mancozeb.

4. *Sclerotinia* Stem Rot

Symptoms

» The stems develop water-soaked spots which later may be covered with a cottony white growth.

» As the disease progresses, aected portions of the stem develop a bleached appearance, and eventually the tissues shred.

» Girdling of the stem results in premature ripening and lodging of plants.

» Hard black bodies, the sclerotia, are formed inside the stem and occasionally on the stem surface.

» Basal stalk infections rarely occur.

Fungus: *Sclerotinia sclerotiorum* (Lib.) de Bary
Syn: *Sclerotinia libertiana* Fuckel; *Whetzelinia sclerotiorum* [Lib.] Korf and Dumont.

Diseases of Mustard

Management

 » Use crop rotation; do not plant highly susceptible crops more than once in four years, including dry edible beans, sunfowers, mustard and canola.

 » Soil application of biocontrol agent *viz., Pseudomonas fluorescens, P. chlororaphis, Streptomyces spiroverticillatus, S. longisporoflavus Coniothyrium minitans* and *Bacillus amyloliquefaciens.*

 » Use at least a five year rotation for severely infested fields.

 » Control broad-leaved weeds.

 » Spray with carbendazim at 0.25%

5. Sclerotium foot rot (Southern blight) and Wilt

Symptoms

 » The infection usually starts at the collar region.

 » Whitish cottony mycelium is seen on the stem and roots.

 » The stem portion shows rotting of tissues at the point of attack and the plants show dropping of leaves and withering finally dry up.

Fungus: *Sclerotium rolfsii* Sacc. (Teleomorph: *Athelia rolfsii* (Curzi) Tu & Kimbrough) The mycelium, hyaline, septate and branched mustard like sclerotia.

Disease cycle: Primary and secondary through sclerotia from infected plant debris as well as from soil

Management

 » Soil application of biocontrol agent *viz., Pseudomonas fluorescens, Trichoderma asperellum* and *Bacillus amyloliquefaciens.*

6. Downy mildew

Symptoms

 » Grayish white irregular necrotic patches develop on the lower surface of leaves.

» Later under favourable conditions brownish white fungal growth may also be seen on the spots.

» The most conspicuous and pronounced symptom is the infection of inflorescence causing hypertrophy of the peduncle of inflorescence and develop stag head structure.

Chromista: *Hyaloperonospora parasitica* (Pers.: Fr.) Fr.

Mycelium is hyaline, coenocytic, remains intercellular in host, produces large, lobed intracellular haustoria, often branched, which nearly fill the entire cell. Erect conidiophores singly or in groups of determinate growth emerge vertically through the epidermis on the undersurface of the leaves through the stomata.

Management

» Seed treatment with metalaxyl 35 SD at 6 g/kg and spray of metalaxyl at 0.01%.

7. Powdery mildew

Symptoms

» Symptoms appear as dirty white, circular, floury patches on either sides of the leaves.

» Under favourable environmental conditions, entire leaves, stems, floral parts and pods are affected.

» The whole leaf may be covered with powdery mass.

Fungus: *Erysiphe cruciferarum* Opiz ex. Junell.

Management

» Foliar spray of carbendazim at 0.25%.

» Dinocap or tridemorph (0.1%) could be sprayed thrice on leaves.

» Spray of 0.04% tridemorph followed by 0.05% hexaconazole, 0.05% tebuconazole, and 0.20% wettable sulfur on leaves was found effective against this disease.

Diseases of Mustard

8.   Black leg or Stem canker

Symptoms

» Blackleg disease causes two distinct types of symptoms, namely, leaf lesions and stem canker.

» Stem cankering is the major reason of yield loss associated with blackleg.

» Symptoms appear first as water-soaked lesions on cotyledons, hypocotyls, and leaves of the host.

» These lesions turn white to gray color, round to irregular in shape, and become dotted with numerous pinhead-sized black asexual fruiting bodies called pycnidia.

» Typical lesions of blackleg can also occur on pods.

Fungus: Teleomorph: *Leptosphaeria maculans* (Desm.) Ces. et de Not.
Anamorph: *Phoma lingam* Tode ex. Fr.

Management

» Spraying of *Trichoderma asperellum*

» Triazole-based fungicides, namely, metconazole, protioconazole, tebuconazole, and flusilazole, were most efficient, and mixture of protioconazole and tebuconazole or flusilazole and carbendazim were very active.

9.   White blister

Symptoms

Local infection

» Isolated pustules or sori develop in leaves and stems.

» Pustules merge to form larger patches, host epidermis rupture after maturity of pustules

Systemic infection

» When young stems and flowering parts are infected it becomes systemic.

» Stimulates hypertrophy and hyperplasia results in enlarged and variously distorted organ mostly flower parts sepals become enlarged to several times.

Chromista: *Albugo candida* (Pers.) Roussel

Mycelium is aseptate, intercellular with nuclei-free globular haustoria.

Management

» Grow resistant varieties like T 4 and YRT 3.

» Combination of metalaxyl 35 ES 6 ml/ kg seed treatment + 0.2 g/ lit spray of combination of metalaxyl + mancozeb at 50 and 65 DAS.

10. Club root

Symptoms

» At the initial stages the affected plants show normal healthy growth, but as the disease develops, the plants become stunted showing pale green or yellowish leaves.

» The plant is then killed within a short time.

» When the plants are pulled, overgrowth (hypertrophy/hyperplasia) of the main and lateral roots becomes visible in the form of small or spindle or spherical-shaped knobs, called clubs.

» Depending on the type of root of a species, the shape of the club varies.

» When many infections occur close together, the root system is transformed into various-shaped malformations.

» The swollen roots contain large numbers of resting spores and plasmodia.

» The older, more particularly the larger, clubbed roots disintegrate before the end of the season.

Protozoa: *Plasmodiophora brassicae* Woronin

Management

» Grow resistant variety like Kalyan (WBBN-1).

» Use of boron 10–30 mg/ kg and calcium nitrate in soil of pH 6.5 or 7.3 was effective in reducing clubroot severity.

» Soil drenching with lime or basic slag 2500 kg/ ha for neutralize the soil.

Diseases of Mustard

» Soil amendment with neem or mahua cake 1.5 ton./ ha.

» Mixing or pouring of some antagonists, namely, *Serratia* spp. and *Trichoderma* spp., by using mushroom compost as a carrier for the antagonists was found effective in reduction of infection.

» Although certain chemicals like azoxystrobin, benomyl, fluazinam, flusulfamide, methyl thiophanate, quintozene, limestone, and other soil fumigants are known to be effective.

11. Bacterial black rot

Symptoms

» Symptoms appear when the plants are 2 months old.

» In the initial stages, dark streaks of varying length are observed either near the base of the stems or 8-10 cm above the ground level.

» Lower leaves show the symptoms first, which include midrib cracking and browning of the veins; when extensive, it brings about withering of the leaves.

» Profuse exudation of yellowish fluid from affected stems and leaves may also occur.

» Blackened veins and V-shaped necrotic lesions on the leaf margins are surrounded by yellow halos.

» The advanced phases of the disease include lesion enlargement, foliar chlorosis, and death of leaves.

» The disease develops from the lower leaves to the apex, resulting in complete leaf necrosis and defoliation.

» The affected plants, on stripping, show a dark brown crust full of bacterial ooze.

» The black rot does not cause any disagreeable odor.

Bacterium: *Xanthomonas campestris* (Pammel) pv. *campestris* (Dowson) Dye, *et al.* (1980)

Management

» Spray captafol 0.2% at 20 day intervals or aureomycin (chlorotetracycline) 200 μg/ mL.

12. Phyllody and Aster Yellows

Symptoms

- » The characteristic symptom is the transformation of floral parts into leafy structures.

- » The corolla becomes green and sepaloid.

- » The stamens turn green and become indehiscent.

- » The gynoecium is borne on a distinct gynophore and produces no ovules in the ovary.

Bacterium: *Candidatus* Phytoplasma asteris

Management

- » Spray acepate 0.75g/ lit.

- » Spray application of copper oxychloride is also reported to give a considerable degree of control of the disease.

13. Mosaic

Symptoms appear as vein clearing, green vein banding, mottling, and severe puckering of the leaves. The affected plants remain stunted and do not produce flowers, or very few flowers are produced on such plants. When siliquae are formed, they remain poorly filled and show shrivelling, which results in decrease in yield and oil content.

Viruses: *Cauliflower mosaic virus*
Rai mosaic virus
Turnip mosaic virus

Chapter - 33

Diseases of Niger
- *Guizotia abyssinica* (Linn. F.) Cass

S. No.	Disease	Pathogen
1.	*Ozonium* wilt	*Ozonium texanum* var. *parasiticum*
2.	Collar rot	*Sclerotium rolfsii*
3.	*Macrophomina* root and stem rot	*Macrophomina phaseolina*
4.	Damping off	*Rhizoctonia solani*
5.	*Cercospora* leaf spot	*Cercospora guizoticola*
6.	*Alternaria* blight	*Alternaria alternata*
7.	Powdery mildew	*Erysiphe cichoracearum*
8.	Rust	*Puccinia guizotiae*
9.	*Curvularia* leaf spot	*Curvularia lunata*
10.	Bacterial leaf spot	*Xanthomonas campestris* pv. *guizotiae*
11.	Cuscuta	*Cuscuta hyalina*

1. *Ozonium* wilt

The disease was reported around Varanasi where it appeared in great severity. Heavy losses are incurred as the diseased plants dry.

296 Diseases of Field Crops and their Management

Symptoms

- » Necrotic lesions develop on stem of well grown plants near the soil level, which gradually extends upwords.

- » Whitish fungal mycelium grows on these necrotic areas under high humidity and the branches, leaves, inflorescence etc become soft and start rotting.

- » Finally, the diseased stem breaks.

Fungus: *Ozonium texanum* var. *parasiticum* Thirum.

The fungus forms rhizomorphs and sclerotia. The sclerotia are oval, grey to dark brown. The recurrence takes place through soil borne sclerotia and mycelium.

Management

- » Seed treatmet with thiram at 2 g/ kg.

2. Collar rot

The disease was first reported from Dharwad, Karnataka.

Symptoms

- » The tissues of collar region become soft and depressed.

- » White fungus grows on the diseased part and forms mustard seed like sclerotia.

- » The diseased plants turn yellow and dry.

Fungus: *Sclerotium rolfsii* Sacc.

- » The disease is soil borne.

Management

- » Deep summer ploughing

- » Soil drenching with copper oxychloride 0.25%.

Diseases of Niger

3. *Macrophomina* root and stem rot

Symptoms

- » Typical root rot, stem rot, charcoal rot and leaf blight symptoms are produced.
- » *Macrophomina* infected roots are light blackish to black in colour, which are covered with black sclerotia and are brittle.
- » The blackening extends from ground level upward on the stem giving black colour to stem.

Fungus: *Macrophomina phaseolina* (Tassi) Goidanich

The recurrence of the disease takes place through seed and soil borne inoculum, later spread is through workers, tools and insects.

Management

- » Deep ploughing in summer
- » Crop rotation
- » Seed treatment with thiram 0.2% + carbendazim 0.1%
- » Application of *Trichoderma viride* at 2.5 kg/ ha mixing with 50 kg FYM in the field before sowing can effectively manage the disease.

4. Damping off/ root rot

Symptoms

- » The fungus attacks stem of the seedling at ground level, makes water soaked soft and incapable of supporting the seedling which falls over and dies.
- » On older seedlings elongated brownish black lesions appear
- » Lesions increase in length and width girdling the stem often the roots are blackened due to fungal attack resulting in root rot.

Fungus: *Rhizoctonia solani* Sacc.

- » The disease is favoured by cold and wet weather.

Management

- » Crop rotation should be followed.
- » Seed treatment with thiram or captan at 3.0 g/ kg seed.
- » Soil drenching with captan 50 WP at 0.25%

5. *Cercospora* leaf spot

Symptoms

- » Disease appears as small straw to brown coloured spots with grey centre on the leaves, spots may coalesce causing defoliation.
- » Later the spots increase in number and size and cover the entire lamina and the leaves start dropping off.
- » Elongated dark brown spots are produced on the stem.
- » The capsules are also affected.

Fungus: *Cercospora guizoticola* Govindu & Thirum.

The pathogen is seed and soil borne. The primary infection occurs through seed borne inoculum as well as the inoculum present on diseased plant left over in the soil. The spread of the disease is through conidia formed on infected plant parts. The pathogen remains active on other collateral hosts also.

Management

- » Seed treatment with carbendazim 50WP 0.2% followed by two foliar sprays of carbendazim 50WP 0.2% + mancozeb 0.1%.

6. *Alternaria* blight

Symptoms

- » The disease appears as concentric rings on the leaves, which turns brown with grey centre later on.
- » As the disease advances, the spots become oval or circular and irregular in shape.

Diseases of Niger

» The infected leaves become dry and lead to the defoliation.

» Further it, spreads to other plant parts and results in to premature drying of the plant

Fungus: *Alternaria alternata* (Fr.) Keissler.

The pathogen is seed and soil borne.

Management

» Use of resistant varieties like JNC 6, IGP 76, Deomali, GA 11 and ONS 8.

» Seed treatment with thiram 0.2% + carbendazim 50WP 0.1%

» Spray with mancozeb 0.25% + carbendazim 50WP 0.1% at 15 days interval, two sprays with zineb or dithane M-45 0.3%.

» Spray mancozeb 0.2% at 15 days interval.

7.   Powdery mildew

Symptoms

» All the aerial parts develop symptoms.

» Small cottony spot develops on the leaves which gradually cover the whole lamina.

» Some times on stem a purple ring also develops.

» The diseased leaves turn yellow and drop off.

» The seed formed on diseased plants are small and shriveled.

Fungus: *Oidium* sp. (*Erysiphe cichoracearum* DC)

The pathogen is known to survive through some unknown collateral hosts.

Management

» The disease can be managed by burning the infected plant parts after the harvesting of the crop.

» The disease can also be effectively controlled by spraying with sulfex at the rate of 0.3% as the disease starts appearing.

300   Diseases of Field Crops and their Management

» Spray of wettable sulphur 0.2% or carbendazim 0.1% or hexaconazole 0.15% or myclobutanil 0.1%.

8. Rust

Symptoms

» Brown pustules up to 7 mm in diameter appear on the leaves.

» The lesions consist of densely aggregated brown telio that measure 0.16 to 0.37 mm in diameter.

» The uredosori are formed on the lower leaf surface and the corresponding upper surface becomes chlorotic.

» Later teliospores are formed.

Fungus: *Puccinia guizotiae*

Management

» Spray with mancozeb 0.2% as the disease starts appearing.

9. *Curvularia* leaf spot: *Curvularia lunata* (Wakker) Boedijn

Symptoms

» Small circular to irregular brown to reddish brown spots which later coalesce to form larger spots.

» The leaf turn yellow dries and defoliates.

» The pathogen is seed borne and survives in plant debris.

10. Bacterial leaf spot

Symptoms

» Small brownish spots are formed on leaves which are surrounded by yellow halo.

» Most of the spots are on mainly on the leaf margin.

» The spots increase, coalesce and give appearance of blight.

Diseases of Niger 301

» All the leaves get destroyed.

» Under severe cases, scabby brown needle pricks lesion produced around needle pricks on the stem.

Bacteria: *Xanthomonas campestris* pv. *guizotiae* (Yirgou 1964) Dye 1978; *Xanthomonas campestris* pv. *indicus* Moniz. Syed & Raj.

Management

» Spray copper oxychloride 0.25%.

11. Cuscutta

Symptoms

» The plants remain stunted, pale yellow and bear a very small number of flowers and fruits due to the association of Cuscuta.

» It is a total stem parasite depending for shelter and food totally on the host.

» The food is obtained through haustoria's from the host usually the Cuscuta seed gets mixed with Niger seed and such mixtures are planted.

Parasite: *Cuscuta hyalina* Roth

Management

» The cuscuta seeds can be removed by sieving before sowing.

» Removal of cuscuta infected niger seed at the early crop growth.

» Apply of thiobencarb at 0.75 kg/ ha and anilofos at 0.50 kg/ ha.

» Pre-emergence application of oxadiazon at 0.75 kg/ha, pendimethalin at 1.0 kg/ ha and alachlor at 1.0 kg/ ha.

» Pre sowing application of fluchloralin at 1 kg a.i./ha or pre emergence soil application of pendimethelin at 1-1.25 kg a.i. /ha.

Chapter - 34

Diseases of Cotton
Gossypium hirsutum, Gossypium barbadense,
Gossypium arboreum, Gossypium herbaceum

S. No.	Disease	Pathogen
1.	*Fusarium* wilt	*Fusarium oxysporum* f. sp. *vasinfectum*
2.	*Verticillium* wilt	*Verticillium dahliae*
3.	Root rot	*Rhizoctonia solani*
4.	Anthracnose	*Colletotrichum capsici*
5.	Grey or areolate mildew	*Ramularia areola*
6.	Boll rot	Fungal complex
7.	Leaf blight	*Alternaria macrospora*
8.	Bacterial blight	*Xanthomonas axonopodis* pv. *malvacearum*
9.	Leaf curl	*Cotton leaf curl virus*
10.	Phyllody (Stenosis)	*Candidatus* Phytoplasma asteris

1. *Fusarium* wilt

In India the disease was reported from Nagpur in 1908 and is retricted to black cotton soil which aer heavy clay with pH 7.6-8.0.

Diseases of Cotton

Symptoms

» The disease affects the crop at all stages.

» The earliest symptoms appear on the seedlings in the cotyledons which turn yellow and then brown.

» The base of petiole shows brown ring, followed by wilting and drying of the seedlings.

» In young and grown up plants, the first symptom is yellowing of edges of leaves and area around the veins i.e. discoloration starts from the margin and spreads towards the midrib.

» The leaves loose their turgidity, gradually turn brown, droop and finally drop off.

» Symptoms start from the older leaves at the base, followed by younger ones towards the top, finally involving the branches and the whole plant.

» The defoliation or wilting may be complete leaving the stem alone standing in the field.

» Sometimes partial wilting occurs; where in only one portion of the plant is affected, the other remaining free.

» The taproot is usually stunted with less abundant laterals.

» Browning or blackening of vascular tissues is the other important symptom, black streaks or stripes may be seen extending upwards to the branches and downwards to lateral roots.

» In severe cases, discolouration may extend throughout the plant starting from roots extending to stem, leaves and even bolls.

» In transverse section, discoloured ring is seen in the woody tissues of stem.

» The plants affected later in the season are stunted with fewer bolls which are very small and open before they mature.

Fungus: *Fusarium oxysporum* f. sp. *vasinfectum* (G.F. Atkinson) Snyder & H.N. Hansen

The mycelium is white to greyish and both inter- intracellular. Macroconidia are 1 to 5 septate, hyaline, thin walled, falcate with tappering ends. The microconidia are hyaline, thin walled, spherical or elliptical, single or two celled. Chlamydospores are dark coloured

304   Diseases of Field Crops and their Management

and thick walled. The fungus also produces a vivotoxin, Fusaric acid which is partially responsible for wilting of the plants.

Disease cycle

» The fungus can survive in soil as saprophyte for many years and chlamydospores act as resting spores. The pathogen is both externally and internally seed-borne.

» The primary infection is mainly from dormant hyphae and chlamydospores in the soil.

» The secondary spread is through conidia and chlamydospores which are disseminated by wind and irrigation water.

Favourable conditions

» Soil temperature of 20-30°C

» Hot and dry periods followed by rains

» Heavy black soils with an alkaline reaction

» Increased doses of nitrogen and phosphatic fertilizers

» Wounds caused by nematode *Meloidogyne incognita* and grubs of Ash weevil (*Myllocerus pustulatus*).

Management

» Grow disease resistant varieties of *G. hirsutum* and *G. barbadense* like Varalakshmi, Vijay, Pratap, Jayadhar, Jerila, Bijoy, BDS and Verum.

» Treat the acid delinted seeds with carboxin or carbendazim at 2 g/ kg.

» Remove and burn the infected plant debris in the soil after deep summer ploughing during June-July.

» Apply increased doses of potash with a balanced dose of nitrogenous and phosphatic fertilizers.

» Apply heavy doses of farm yard manure or other organic manures.

» Follow mixed cropping with non-host plants.

» Spot drench with carbendazim 1g/ litre.

Diseases of Cotton

2. *Verticillium* wilt

Symptoms

- » The symptoms are seen when the crop is in squares and bolls.
- » Plants infected at early stages are severely stunted.
- » The first symptoms can be seen as bronzing of veins.
- » It is followed by interveinal chlorosis and yellowing of leaves.
- » Finally the leaves begin to dry, giving a scorched appearence.
- » At this stage, the characteristic diagnostic feature is the drying of the leaf margins and areas between veins, which gives a "Tiger stripe" or "Tiger claw" appearance.
- » The affected leaves fall off leaving the branches barren. Infected stem and roots, when split open, show a pinkish discolouration of the woody tissue which may taper off into longitudinal streaks in the upper parts and branches.
- » The infected leaf also shows brown spots at the end of the petioles.
- » The affected plants may bear a few smaller bolls with immature lint.

Fungus: *Verticillium dahlia* Kleb.

The fungus produces hyaline, septate mycelium and two types of spores. The conidia are single celled, hyaline, spherical to oval, borne singly on verticillate condiophores. The micro sclerotia are globose to oblong, measuring 48-120×26-45 μm.

Disease cycle

The fungus also infects the other hosts like brinjal, chilli, tobacco and bhendi. The fungus can survive in the infected plant debris and in soils as micro sclerotia upto 14 years. The seeds also carry the micro sclerotia and conidia in the fuzz. The primary spread is through the micro sclerotia or conidia in the soil. The secondary spread is through the contact of diseased roots to healthy ones and through dissemination of infected plant parts through irrigation water and other implements.

Favourable conditions

- » Low temperature of 15-20°C.

306   Diseases of Field Crops and their Management

- » Low lying and ill-drained soils.

- » Heavy soils with alkaline reaction.

- » Heavy doses of nitrogenous fertilizers.

Management

- » Grow disease resistant varieties like Sujatha, Suvin and CBS 156 and tolerant variety like MCU 5, WT.

- » Treat the delinted seeds with carboxin or carbendazim at 2 g/ kg.

- » Remove and destroy the infected plant debris after deep ploughing in summer months (June-July).

- » Apply heavy doses of farmy and manure or compost at 100 t/ ha.

- » Follow crop rotation by growing paddy or lucerne or chrysanthemum for 2-3 years.

- » Spot drench with carbendazim 0.2%.

3.   Root rot

Symptoms

- » The pathogen causes three types of symptoms *viz.*, seedling disease, sore-shin and root rot.

- » Germinating seedlings and seedlings of one to two weeks old are attacked by the fungus at the hypocotyl and cause black lesions, girdling of stem and death of the seedling, causing large gaps in the field.

- » In sore-shin stage (4 to 6 weeks old plants), dark reddish-brown cankers are formed on the stems near the soil surface, later turning dark black and plant breaks at the collar region leading to drying of the leaves and subsequently the entire plant.

- » Typical root rot symptom appears normally at the time of maturity of the plants.

- » Initially, all the leaves droop suddenly and die with in a day or two.

- » The affected plants when pulled reveal the rotting of entire root system except tap root and few laterals.

- » The bark of the affected plant shreds and even extends above ground level.

Diseases of Cotton

» In badly affected plants the woody portions may become black and brittle.

» A large number of dark brown sclerotia are seen on the wood or on the shredded bark.

Fungus: *Rhizoctonia solani* Kuhn.

The fungal hyphae are septate and fairly thick and produce black, irregular sclerotia which measure 100 m in diameter.

Disease cycle

The disease is mainly soil-borne and the pathogen can survive in the soil as sclerotia for several years. The spread is through sclerotia which are disseminated by irrigation water, implements, and other cultural operations.

Favourable conditions

» Dry weather following heavy rains,

» High soil temperature (35-39°C),

» Cultivation of favourable hosts like vegetables,

» Oil seeds and legumes preceding cotton

» Wounds caused by ash weevil grubs and nematodes.

Management

» Seed treatment with *Trichoderma asperellum* at 4 g/ kg.

» Spot drench with carbendazim 0.1%.

» Apply farm yard manure at 10t/ha or neem cake at 150 kg/ ha.

» Adjust the sowing time, early sowing (First Week of April) or late sowing (Last week of June) so that crop escapes the high soil temperature conditions.

» Adopt intercropping with sorghum or moth bean (*Phaseolus aconitifolius*) to lower the soil temperature.

4. Anthracnose

Symptoms

» The pathogen infects the seedlings and produces small reddish circular spots on the cotyledons and primary leaves.

» The lesions develop on the collar region, stem may be girdled, causing seedling to wilt and die.

» In mature plants, the fungus attacks the stem, leading to stem splitting and shredding of bark.

» The most common symptom is boll spotting.

» Small water soaked, circular, reddish brown depressed spots appear on the bolls.

» The lint is stained to yellow or brown, becomes a solid brittle mass of fibre.

» The infected bolls cease to grow and burst and dry up prematurely.

Fungus: *Colletotrichum capsici* Syd.

The pathogen forms large number of acervuli on the infected parts. The conidiophores are slightly curved, short, and club shaped. The conidia are hyaline and falcate, borne single on the conidiophorous. Numerous black coloured and thick walled setae are also produced in acervulus.

Disease cycle

The pathogen survives as dormant mycelium in the seed or as conidia on the Surface of seeds for about a year. The pathogen also perpetuates on the rotten bolls and other plant debris in the soil. The secondary spread is by air-borne conidia. The pathogen also survives in the weed hosts *viz., Aristolachia bractiata* and *Hibiscus diversifolius.*

Favourable conditions

» Prolonged rainfall at the time of boll formation

» Close planting.

Diseases of Cotton

Management

- » Treat the delinted seeds with carbendazim or carboxin or captan at 2g/ kg.

- » Remove and burn the infected plant debris and bolls in the soil.

- » Rogue out the weed hosts.

- » Spray with mancozeb 2 kg or copper oxychloride 2.5 kg or or carbendazim 500 g/ ha at boll formation stage.

5.   Grey or areolate mildew

Symptoms

- » The disease usually appears on the under surface of the bottom leaves when the crop is nearing maturity.

- » Irregular to angular pale translucent lesions which measure 1-10 mm (usually 3-4 mm) develop on the lower surface, usually bound by vein lets.

- » On the upper surface, the lesions appear as light green or yellow green specks.

- » A frosty or whitish grey powdery growth, consisting of conidiophores of the fungus, appears on the lower surface.

- » When several spots coalesce, the entire leaf surface is covered by white to grey powdery growth.

- » White or grey powdery growth may occur on the upper surface also.

- » The infection spreads to upper leaves and entire plant may be affected.

- » The affected leaves dry up from margin, cup inward; turn yellowish brown and fall of prematurely.

Fungus: Teleomorph: *Mycosphaerella areola* Ehrlich & F.A. Wolf
Anamorph: *Ramularia areola* G.F. Atk.

The pathogen produces endophytic, septate mycelium. Conidiophorous are short, hyalineand branched at the base. Conidia are borne singly or in chains at the tips of conidiophores. The conidia are hyaline, irregularly oblong with pointed ends, sometimes rounded to flattend ends, unicellular or 1-3 septate. The perfect stage of the fungus produces perithecia containing many asci. The ascospores are hyaline and usually two celled.

310

Disease cycle

The pathogen survives during the summer in the infected crop residues. The perennial cotton plants and self-sown cotton plants also harbour the pathogen during summer months. The primary infection is through conidia from infected plant debris and secondary spread is through wind, rain splash, irrigation water and implements.

Favourable conditions

> » Wet humid conditions during winter cotton season,

> » Intermittent rains during North-East monsoon season,

> » Low temperature (20-30°C) during October-January,

> » Close planting, excessive application of nitrogenous fertilizers,

> » Very early sowing or very late sowing of cotton

Management

> » Remove and burn the infected crop residues.

> » Rogue out the self-sown cotton plants during summer months.

> » Avoid excessive application of nitrogenous fertilizers/manures.

> » Adopt the correct spacing based on soil conditions and varieties.

> » Grow resistant varieties like Sujatha, Varalakshmi and Savitri.

> » Spray the crop with carbendazim at 500 g/ ha, repeat after a week.

6. Boll rot - Fungal complex

It is a complex disease caused by several fungal pathogens *viz., Fusarium moniliforme, Colletotrichum capsici, Aspergillus flavus, A. niger, Rhizopus nigricans, Nematospora nagpuri* and *Botryodiplodia* sp.

Symptoms

Initially, the disease appears as small brown or black dots which later enlarge to cover the entire bolls. Infection spreads to inner tissues and rotting of seeds and lint occur. The bolls never burst open and fall off and prematurely. In some cases, the rotting may be external,

Diseases of Cotton

causing rotting of the pericarp leaving the internal tissues free. On the affected bolls, a large number of fruiting bodies of fungi are observed depending upon the nature of the fungi involved.

Fusarium moniliforme	*Aspergillus flavus*
Mycelium: Septate, hyaline, Conidiophore-slender, short, hyaline, simple, stout or branched irregularly. **Two types of conidia**: macroconidia (several celled, slightly curved or bent, pointed at the both the ends, sickle shaped), microconidia (1 or 2 celled, ovoid, single or in chains, hyaline). **Chlamydospores**: terminal or intercalary, produced singly or in chains by the mycelial hyphae or macroconidia.	**Mycelium**: highly branched, septate. Conidiophore: characteristic symmetric or asymmetric broom like fashion. The first generation branches are called primary branches or rammi, on which whorls of second generation branches called metulae are produced. Each metula ultimately bears bottled shaped phialides which bears conidia in chains in basipetal succession. **Conidia**: globose, hyaline.

Disease cycle

The fungi survive in the infected bolls in the soil. The insects mainly help in the spread of the disease. The fungi make their entry only through wounds caused by the insects. The secondary spread of the disease is also through air-borne conidia.

Favourable conditions

> Heavy rainfall during the square and boll formation stage,

> Wounds caused by the insects,

> Especially red cotton bug *Dysdercus cingulata*

> Close spacing and excessive nitrogen application.

Management

> Adopt optimum spacing.

> Apply the recommended doses of fertilizers.

> Spray copper oxychloride 2.5 kg along with an insecticide for bollworm from 45th day at 15 days interval, two or three sprays are necessary.

7. Leaf blight

Symptoms

- » The disease may occur in all stages but more severe when plants are 45-60 days old.
- » Small, plate to brown, irregular or round spots, measuring 0.5 to 6mm diameter, may appear on the leaves.
- » Each spot has a central lesion surrounded by concentric rings.
- » Several spots coalesce together to form blighted areas.
- » The affected leaves become brittle and fall off.
- » Sometimes stem lesions are also seen.
- » In severe cases, the spots may appear on bracts and bolls.

Fungus: *Alternaria macrospora* Zimm.

The fungus produces dark brown, short, 1-8 septate, irregularly bend conidiophores with a single conidium at the apex. The conidia are obclavate, light to dark brown in colour with 3-9 transverse septa and four longitudinal septa, with a prominent beak.

Disease cycle

The pathogen survives in the dead leaves as dormant mycelium. The pathogen primarily spreads through irrigation water. The secondary spread is mainly by airborne conidia.

Favourable conditions

- » High humidity.
- » Intermittent rains.
- » Moderate temperature of 25-28°C

Management

- » Remove and destroy the infected plant residues.
- » Grow resistant varieties like K7, K8, MCU 5 and Jayadhar.

Diseases of Cotton

» Spray mancozeb 2 kg or copper oxychloride at 2 kg/ ha at the intimation of the disease, four to five sprays may be given at 15 days interval.

8. Bacterial blight

» First reported in Alabama province of USA by Atkinson (1891).
» In India first reported at Madras (1918) and only occurs on exotic tetraploid (*Gossypium hirsutum* and *Gossypium barbadense*).

Symptoms

The bacterium attacks all stages from seed to harvest. Usually five common phases of symptoms are noticed.

i. Seedling blight:

Small, water-soaked, circular or irregular lesions develop on the cotyledons, later, the infection spreads to stem through petiole and cause withering and death of seedlings.

ii. Angular leaf spot:

Small, dark green, water soaked areas develop on lower surface of leaves, enlarge gradually and become angular when restricted by veins and veinlets and spots are visible on both the surface of leaves. As the lesions become older, they turn to reddish brown colour and infection spreads to veins and veinlets.

iii. Vein blight or vein necrosis or black vein:

The infection of veins cause blackening of the veins and veinlets, gives a typical 'blighting' appearance. On the lower surface of the leaf, bacterial oozes are formed as crusts or scales. The affected leaves become crinkled and twisted inward and show withering. The infection also spreads from veins to petiole and cause blighting leading to defoliation.

iv. Black arm:

On the stem and fruiting branches, dark brown to black lesions are formed, which may girdle the stem and branches to cause premature drooping off of the leaves, cracking of stem and gummosis, resulting in breaking of the stem and hang typically as dry black twig to give a characteristic "black arm" symptom.

314   Diseases of Field Crops and their Management

v. Square rot / Boll rot:

On the bolls, water soaked lesions appear and turn into dark black and sunken irregular spots. The infection slowly spreads to entire boll and shedding occurs. The infection on mature bolls leads to its premature bursting. The bacterium spreads inside the boll and lint gets stained yellow because of bacterial ooze and looses its appearance and market value. The pathogen also infects the seed and causes reduction in size and viability of the seeds.

Bacterium: *Xanthomonas axonopodis* pv. *malvacearum* (E. F. Smith) Vauterin

The bacterium is a short rod with a single polar flagellum. It is Gram negative, non-spore forming and measures 1.0-1.2 × 0.7-0.9 μm.

Disease cycle

The bacterium survives on infected, dried plant debris in soil for several years. The bacterium is also seed-borne and remains in the form of slimy mass on the fuzz of seed coat. The bacterium also attacks other hosts like *Thumbergia thespesioides, Eriodendron anfructuosum, Lochnera pusilla* and *Jatropha curcus.* The primary infection starts mainly from the seed-borne bacterium. The secondary spread of the bacteria may be through wind, wind blown rain splash, irrigation water, insects and other implements. Remain viable in in dried leaves upto 17 years and in soil 8 days.

Favorable conditions

» Optimum soil temperature of 28°C and high atmospheric temperature of 30-40°C.

» Relative humidity of 85%, early sowing, delayed thinning, poor tillage, late irrigation and potassium deficiency in soil.

» Rain followed by bright sunshine during the months of October and November are highly favorable.

Management

» Delint the cotton seeds with concentrated sulphuric acid at 100 ml/ kg.

» Treat the delinted seeds with carboxin or oxycarboxin at 2 g/kg or soak the seeds in 1000 ppm Streptomycin sulphate overnight.

» Presowing of seeds treated with acid delinting with sulphuric acid (200 ml/ kg of seed) followed by seed soaking in streptomycin 100 ppm.

Diseases of Cotton

» Remove and destory the infected plant debris. Rogue out the volunteer cotton plants and weed hosts.

» Follow crop rotation with non-host crops.

» Early thinning and early earthing up with potash.

» Grow resistant varieties like Sujatha, 1412 and CRH 71.

» Spray with streptomycin sulphate + tetracycline mixture 100g along with copper oxychloride at 1.25 kg/ ha.

» Growing of diploid cotton and resistant line *G. herbaceum* var. *punctatum*

» Spray neem based formulations

» Hotwater treatment of seeds at 56°C for 10 min is found eradicating in the seed borne infection.

9. Leaf curl

Symptoms

» Downward and upward curling of leaves and thickening of veins and enation on underside of leaves are the characteristic symptoms of the disease.

» In serve infection all the leaves are curled and growth retarded.

» Boll bearing capacity is reduced

Virus: *Cotton leaf curl virus*

It is caused by *Cotton leaf curl virus* - a begomovirus of family geminiviridae. The virions are typical geminate particles, ss circular DNA, bipartite genome with DNA-A and DNA-B components. In south India this disease was caused by *Tobacco streak virus.*

Disease cycle

The primary source is the viruliferous whitefly vector *Bemisia tabaci*. The alternate hosts and cultivated hosts serve as virus reservoirs throughout the year. Not transmitted by seed or contact.

Management

- » Management of planting date to avoid peak vector population.
- » Elimination of volunteer perennial cotton and alternate hosts including malvaceous hosts like wild okra
- » Use of fungus *Paecilomyces farinosus* which parasitizes *B. tabaci*.
- » Foliar spray of neem leaf extract and neem oil 1% resulted in reduction of virus.
- » Vector management by application of granular systemic insecticides.

10. Phyllody (Stenosis)

Symptoms

- » The disease appears when the plants are two to three months old and affected plants are stunted.
- » They put forth numerous extremely small leaves in cluster and the dormant buds are stimulated resulting in profuse vegetative growth.
- » The leaves are disfigured and variously lobed.
- » Flowers remain small with abortive ovary.
- » Large number of flower buds and young seeds.
- » Root system is poorly developed and can be easily pulled out.
- » Sometimes, the disease affects only the base of the plant, resulting in the formation of clump of short branches which bear small and deformed leaves.
- » The mode of transmission of disease and the role of vector are unknown.

Bacterium: *Candidatus* Phytoplasma asteris

Transmitted by *Orosius cellulosus*

Management

- » Rogue out the infected plants periodically.
- » Cotton varieties developed from *Gossypium hirsutum* and *G. barbadense* are found to be resistant to the disease.

Diseases of Cotton 317

Minor diseases

Leaf spot: *Cercospora gossypina* Cooke

- » Round or irregular grayish spots with dark brown or blackish borders appear on older leaves.

Myrothecium leaf spot: *Myrothecium roridum* Tode ex Fr.

- » Reddish spots of 0.5 mm- 1 cm diameter may appear near the margins of the leaves.
- » The affected portions fall off leaving irregular shot holes in the leaves.

Rust: *Phakopsora desmium* (Arthur) Hirats. f.

- » Yellowish brown raised pustules appear on the lower surface of leaves with rusty spores.
- » Several pustules join to give rusty appearance to entire leaf.
- » The sori may also develop on bolls.

Sooty mould: *Capnodium* sp.

- » Dark specks appear on the leaves and bolls, slowly spread and black powdery growth covers the entire leaf area and bolls.

Chapter - 35

Diseases of Jute: *Corchorus capsularis* L. (White jute), *Corchorus olitorius* L. (Tossa Jute)

S. No.	Disease	Pathogen
1.	Root and stem rot	*Macrophomina phaseolina*
2.	Anthracnose	*Colletotrichum corchori*
3.	Black band	*Botryodiplodia theobromae*
4.	Soft rot	*Sclerotium rolfsii*
5.	Powdery mildew	*Oidium* sp.
6.	Hooghly wilt	Bacterial-Fungal-Nematode Complex
7.	Leaf mosaic	*Jute leaf mosaic virus*

1. Root and stem rot

Symptoms

- » Leaf turns pale gray colour in the mid rib and turns black.

- » The spot gradually grows lengthwise, encircles the stem, which internally rots and breaks the plant.

- » As a result the plants die.

- » The major disease of jute, stem rot, initiates at the seedling stage, when the plant height become 6 to 8 inches and it take place till adult stage of jute plant.

Diseases of Jute

» Brownish spot is noticeable on the leaves.

» These spots may be seen from the lower level to the apex of the plant.

» Black dots are present at the brownish pretentious place.

» Eventually the precious consign break down resulting in death of plant.

» Disease is disseminated by seed, soil and air.

» Deshi and Tossa jute are infected by this disease.

Fungus: *Macrophomina phaseolina* (Tassi) Goid.

The pycnidial and sclerotial stages are responsible for the disease and the perfect stage, *Orbilia obscura* is very rarely seen. The pycnida are initially immersed in host tissues, then erumpent at maturity. The pycnidia bear simple, rodshaped conidiophores. Conidia are single celled, hyaline, and elliptic or oval. Micro-sclerotia are formed from aggregates of hyphal cells joined by a melanin.

Disease cycle

The sclerotium survives in the soil and on infected crop debris (upto three years) serve as the primary source of inoculum. Sclerotia germinate on the root surface, germ tubes form appresoria that penetrate the host epidermal cell walls by mechanical pressure and enzymatic digestion or through natural openings.

Favourable conditions

» Alluvial and lateritic soils with low pH (5.6-6.5), high level of nitrogen, high rainfall and high humidity favour infection of *M. phaseolina*.

» Higher soil temperature and low soil moisture predispose the older plants.

» Continuous cultivation of jute crop in the same field may cause depletion of calcium, potassium and other basic elements and make the soil acidic which favours the incidence of stem rot disease.

Management

» Burn the crop debris.

» Crop rotation with non-host crops like rice, wheat, mustard etc.

320 Diseases of Field Crops and their Management

» Proper sanitation reduces the possibility of the disease.

» Generally, lime or dolomite at 2-4 t/ ha is applied about one month before sowing for correction of soil pH.

» Grow resistant lines of *C. capsularis* like JRC 212, JRC 321, JRC 747, D 154, CIM 036, CIM 064, CIN 109, CIN 362, CIN 360 and CIN 386.

» Grow moderately resistant lines of *C. olitorius* like JRO 632, JRO 878, JRO 7835, OIN 125, OIN 154, OIN 651 and OIN 853

» Application of potash (K_2O at 50-100 kg/ ha) reduces the disease severity.

» Seed treatment with powder formulation of *T. viride* at 10 g/ kg at final ploughing and its soil application (by mixing at 1 kg formulation in 100 kg of FYM, cover it for 7 days with polythene, sprinkle water and turn the mixture in every 3-4 days interval and then broadcast the mixture in the field at the time of final ploughing).

» Seed treatment with carbendazim 50 WP at 2 g/ kg or mancozeb at 5g/kg.

» Spray of carbendazim 50 WP at 2 g/ lit or copper oxychloride 50 WP at 5-7 g/ lit or tebuconazole 25.9 EC at 0.1% or mancozeb at 2g/ lit or validamycin 2.5 ml/ lit at 2-3 times at the plant base soil.

2. Anthracnose

Symptoms

» At seedling stage, the disease appeared on leaf and stem as brownish spot and streaks followed by drying up.

» On mature plants, initially light yellowish patches are seen on stem which turns to brown/black depressed spots.

» The spots are irregular in shape and size.

» Several spots may coalesce causing deep necrosis showing crakes on the stem and exposing the fibre tissues.

» The coalescing spots may often girdle the stem the plants break at that point and die.

» Affected plants when survived show necrotic wounds all over the stem.

» Fibres extracted from such affected plants are specky and knotty and fall under very low grade (cross bottom).

Diseases of Jute

» Pods of diseased plants are also affected showing depressed spots and seeds collected from such fruits are also infected.

» The infected seeds are lighter in colour, shrunken and germination is poor.

Fungi: *Colletotrichum corchori* Ikata & Yoshida in white jute
Colletotrichum gloeosporioides (Penz.) Penz. & Sacc. in tossa jute

Disease cycle

The pathogen survives in the soil, infected crop debris and in seeds. The mycelium enters through the epidermis and attacks the parenchymatous tissues between the wedges of bast fibre bundles. The entire parenchymatous tissue of cortex gradually disintegrates.

Favourable conditions

» Continuous rain, high relative humidity and temperature of around 35°C.

Management

» Removal of affected plants and clean cultivation reduce the disease.

» Seed treatment with carbendazim 50 WP at 2 g/ kg or captan at 5 g/ kg of seed

» Crop rotation with rice and wheat etc.

» Spray of carbendazim 50 WP at 2 g/ lit or captan at 5 g/ lit or mancozeb at 5 g/lit.

3.   Black band

Symptoms

» The lesion first appears as small black blackish brown patch, which gradually enlarges and encircles the stem making a black band around.

» Disease disseminated by seed, soil and air. Deshi and Tossa jute infected by this disease.

» On rubbing the stem surface, unlike stem rot profuse black shooty mass of spores adhere to the fingers.

» This disease is seed, soil and air borne.

322   Diseases of Field Crops and their Management

Fungus: *Lasiodiplodia theobromae* Pat.

Management

- » Clean cultivation

- » Seed treatment with carbendazim 50 WP at 2 g/ kg and foliar application of carbendazim 50 WP at 2 g/ lit or copper-oxychloride at 5-7 g/ lit or mancozeb at 4-5 g/ lit.

4.   Soft rot

Symptoms

- » On stem, soft brown wet patch is observed at the point of infection.

- » The bark peels off and the exposed fibre layers turns rusty brown and the affected plants wilt.

- » White mycelial growth and brown globose to sub-globose mustard like sclerotia are also observed at the site of infection.

Fungus: *Athelia rolfsii* (Curzi) C.C. Tu & Kimbr.
The pathogen is soil borne and having a large number of hosts.

High soil moisture with high temperature and close spacing favours the disease.

Management

- » Clean and weed free cultivation.

- » Summer ploughing

- » Spray copper-oxychloride at 5-7 g/ lit of water at base region reduces the disease.

5.   Powdery mildew

Symptom

- » At the end of the jute season fine white powdery mass appears to be accumulated on leaf-surface resulting fall of leaves, flower and fruits.

- » Foggy weather is favourable for the growth.

Diseases of Jute

» Disease disseminated by seed, soil and air.

» Deshi and Tossa jute infected by this disease.

Fungus: *Oidium* sp.

Management

» Spray sulphur at 3.2 g/ lit.

6.   Hooghly wilt

Symptoms

» Root system of affected plant becomes infested with a soil borne fungi.

» All the leaves become flaccid at a time and after few days dropping occurs.

» At the flowering stage, wilting occurs severely on jute plants.

» *Rhizoctonia solani* silted the whole plant and it becomes dried off which show the way to death of the plant.

» The *Olitorius* varieties are affected by this disease more than the *Capsularis*.

» Disease disseminated by seed and soil.

» Toss jute is infected by this disease.

Bacterial-Fungal-Nematode Complex: *Ralstonia solanaearum* (Smith) Yabuuchi *et al.*, *Meloidogyne incognita, Rhizoctonia bataticola* (Taubenh.) E.J.Butler and *Fusarium* complex

Ralstonia solanacearum (=*Pseudomonas solanacearum*) is primary pathogen and *R. bataticola* and *Meloidogyne incognita* are associated pathogens. Presence of these root rot pathogen (*R. bataticola*/ *M. phaseolina*) and root knot nematode (*M. incognita*) increases the disease since they create wound in the root, and thus facilitate the entry of the primary bacterial pathogen, *R. solanacearum*.

Management

» Clean cultivation

» Common hosts like potato or other solanaceaous crops in rotation with jute are to be avoided.

324 Diseases of Field Crops and their Management

- » Jute: Paddy: Paddy or Jute: Paddy: Wheat are the most effective rotation.

- » Removing wilt affected plants from the field, burning of the dead plants and rotten potato tubers are important cultural practices to keep the disease under control.

- » Seed treatment with carbendazim 50 WP at 2 g/ kg.

- » Spary carbendazim 50 WP at 2 g/ lit or copper-oxychloride at 5-7 g/ lit or mancozeb at 4-5 g/ lit.

7.   Leaf mosaic

Symptoms

- » Yellow mosaic spots regular or irregular appear usually on *Capsularis* plants at any stage of growth affecting formation of chlorophyll.

- » Leaves in some cases produce small enation along the mid vein.

- » In extreme cases the infected plant gets stunted and leads to reduced plant height.

Virus: *Jute leaf mosaic virus*

It is transmitted by white fly (*Bemisia tabaci*)

Management

- » Uprooting of infected plants

- » Spray imidacloprid 17.8 SL at 0.3%

Minor diseases

- » **Stem gall**: *Protomyces* sp.

- » **Tip blight**: *Curvularia subulata* (Nees ex Fr.) Boedijn ex J.C. Gilman

- » *Corchorus yellow vein virus*

- » *Corchorus golden mosaic virus*

Chapter - 36

Diseases of Silk-cotton - *Bombax ceiba* L.

1. Brown leaf spot

Symptoms

- » At first, small, round to irregular and yellowish spots, 2-5 mm in diam., appear at the central or marginal area of leaves.

- » They soon enlarge to 5-10 mm in size and sometimes coalesce to each other to form bigger lesions.

- » Then brown necrotic area arises at the center of yellowish spots.

- » Lower surface of the necrotic spots is covered with dark greenish to blackish sooty masses consisting of conidiophores and conidia.

- » The disease is usually prevalent on kapok seedlings and planted young trees.

- » Diseased leaves easily defoliate by shaking them.

- » In case of severely attacked seedlings, only a few leaves survive at the top of shoot or twigs and growth of seedlings is remarkably retarded.

Fungus: *Pseudocercospora italica* Curzi

Management

» Spray mancozeb 0.2% or chlorothanil 0.2%.

Dieback: *Colletotrichum capsici* (Syd.) E.J. Butler & Bisby, *Fusarium solani* (Mart.) Sacc and *Lasiodiplodia theobromae* (Pat.) Griffon & Maubl.

Leaf spots: *Cercospora bombycina* Chidarwar 1382

                    Cercospora ceibae Chupp

                    Corynespora cassicola (Berk. & M.A. Curtis) T. Wei.

                    Myrothecium roridum Tode. ex Fr.

                    Phyllosticta bombacis Batista

                    Phoma holoptelea

Leaf bight: *Ascochyta bombacina*

Seedling blight: *Sclerotium rolfsii* Sacc and *Ascochyta* sp.

Anthracnose: *Colletotrichum dematium* (Pers.) Grove

Sooty mould: *Capnodium brazilienze* Puttemans (1903)

Collar rot: *Rhizoctonia solani*

Root rot: *Armillaria ostoyae, Calonectria* sp. and *Camillea* sp.

Tumour: *Agrobacterium tumefaciens* Smith & Townsend, 1907

Wood stains: *Botryodiplodia theobromae, Ceratocystis fimbriata, Acremonium* sp., *Scytalidium lignicola* and *Fusarium* sp.

Phanerogamic parasite: *Dendrophthoe falcate* (L.f.) Ettingsh

Chapter - 37

Diseases of Sunn hemp
- *Crotalaria juncea* L.

S. No.	Disease	Pathogen
1.	Vascular wilt	*Fusarium udum* f. sp. *crotalariae*
2.	Anthracnose	*Colletotrichum curvatum*
3.	Leaf blight	*Macrophomina phaseolina*
4.	*Choanephora* twig blight	*Choanephora cucurbitarum*
5.	Sunn hemp mosaic	*Sunnhemp mosaic virus*
6.	Southern sunn hemp mosaic	*Tobacco mosaic virus*
7.	Sunn hemp phyllody	*Candidatus* Phytoplasma
8.	Sunn hemp leaf curl	*Sunn hemp leaf curl virus*

1. Vascular wilt

Symptoms

» The affected plant gradually whither, droops, hangs down, turns brown, and ultimately dies within a day or two.

» In grown up plants the wilting parts droop at the tips and defoliation starts which consequently die.

» The discolouration of the tissues could be traced to the main tap root or lateral roots.

Fungus: *Fusarium udum* (Bult) f. sp. *crotalariae* (Kulkarni)

Disease cycle

The fungus survives in the soil as well as in crop residues as facultative parasite. It attacks the plant through the thinner roots and rootlets, even through the cracking in the basal portion of the stem occurs.

Favourable conditions

> It is favoured by high temperature and relative humidity.

Management

> Grow resistant varieties like K-12 (Yellow) and K-12 (Black)

> Seed treatment with carbendazim at 2 g/ kg.

> Spray carbendazim at 2 g/ lit.

> Soil application of neem cake along with seed treatment and application of $ZnSO_4$

2. Anthracnose

Symptoms

> The disease makes an appearance in the form of soft discoloured areas on the cotyledon.

> Later, brownish spots are formed on all parts of host except underground parts.

> The affected cotyledons themselves drop from the petiole.

> The infection spreads downward and acervuli are formed with copious spores on the infected areas within two days. The young seedling when infected generally dies.

> The spots on older leaves appear on one side of the leaf but gradually enlarge and extend to the opposite side.

> These spots are grayish brown to dark brown, roundish or irregular. Several spots coalesce and cover the entire leaf.

Fungi: *Colletotrichum curvatum* Briant and Martyn; *Colletotrichum crotolariae.*

Diseases of Sunn hemp

Disease cycle

The acervuli are formed in the epidermis of the diseased area on which numerous single celled hyaline condia are formed which spread the disease.

Favourable conditions

- » The cloudy weather accompanied by continuous rain favour the rapid spread of disease in the thickly populated crop.
- » The infection is found to be severe in the seedling stage.
- » Rain splashing helps the spores to spread in adjacent plants.

Management

- » Grow tolerant varieties like SH-4, SUN 053, SUIN 037 and JRJ 610
- » Seed treatment with carbendazim at 2 g/ kg.

3. Leaf blight

Symptoms

- » The blight started from the margin of the leaf and proceeds inwards.
- » In the early morning, the blighted leaves look greyish and water soaked, which ultimately become brownish with broad yellow margin.
- » Subsequently, the infected leaf becomes weak and droops down from the plant, which gives a sickly appearance of the whole field.

Fungus: *Macrophomina phaseolina* (Tassi) Goid.

Disease cycle

The black pinhead like pycnidia is noticed on blighted site on which numerous single celled hyaline condia are formed which spread the disease.

Favourable conditions

- » Under moist and warm conditions with intermittent rains.
- » High rainfall and high humidity, early sowing in dry season.

Management

- » Seed treatment with thiram or carbendazim 2 g/ kg followed by spraying with carbendazim 50 WP at 0.2% at 15 days interval.

4. *Choanephora* **twig blight**

Symptoms

- » This disease is characterized by the rotting of terminals.
- » Brown discolouration occurred just below the infection point. Affected portions decay, break and droop.
- » Infected leaves loose their chlorophyll and droop from the stem.
- » The pathogen affects the epidermal and outer cortical layers, and cells in these regions get disintegrated.

Fungus: *Choanephora cucurbitarum* (Berk. & Ravenel) Thaxt.,

Disease cycle

- » White mycelial growth along with black coloured sexual bodies of the fungus is seen on the affected parts from the tip towards petioles.

Favourable conditions

- » High rainfall and humidity favours the disease incidence.

Management

- » Spray with carbendazim 50 WP at 0.2% at 15 days interval.

5. **Sunn hemp mosaic**

Symptoms

- » Mottling appeared within 10 to 12 days after on the youngest leaves.
- » In progress, patches of light and dark green areas became more prominent.
- » Diseased leaves were smaller than the normal.

Diseases of Sunn hemp

» In severe cases of infection, growth of the lamina generally became abnormal.

» Frequently dark green raised areas on the upper surface were seen with corresponding depression on the lower surface.

» The infected plant remained shorter in height and therefore, fibre and seed yields were greatly reduced.

Virus: *Sunn hemp mosaic virus.* The virus consists of spherical particles measuring 26-40µm.

Favourable conditions

» High nitrogen dose.

Management

» Grow tolerant varities like H-4, SUIN 053 and SUIN 037

6. Southern sunn hemp mosaic

Symptoms

» Initially, faint discoloured patches appearing first on young leaves.

» Distinct mosaic with puckering and blistering of leaves developed.

» Subsequently thin elogated enations running more or less parallel to each other developed on the under surface of the leaves.

» The characteristics mosaic mottle and varying degree of leaf distortion occurred in advanced stage of the disease.

» Plants became dwarf with reduced leaves and bear scanty flush of flower resulting in poor pod setting and seed yield.

Virus: *Tobacco mosaic virus.* It is rod shaped with 300µm × 18µm.

Management

» Clean cultivation

7. Sunn hemp phyllody

Symptoms

» It is characterized by yellowing of apical leaf, followed by big bud formation at the terminal raceme, conversion of floral meristem to the vegetative state, leading to buds becoming phylloid and forming dwarf shoot.

Management

» Spray tetracycline at 500 ppm.

8. Sunn hemp leaf curl

Symptoms

» Initially, light faint mosaic mottling and leaf curling.

» But with time the symptoms intensified with severe mosaic, yellowing mottling and leaf curling (upwards and downwards) with reduction in leaf size and plant height.

Virus: *Sunn hemp leaf curl virus*; *Indian tomato leaf curl virus*

Management

» Spray imidaclorprid 17.8 SL at 0.3% or thiamethoxam 25 WG at 0.25%.

Minor diseases

» **Powdery mildew**: *Microsphaera diffusa* Cook and Peck); *Leveillula taurica* Lev.

» **Root and stem rot**: *Sclerotium rolfsii* Sacc.

» **Root rot**: *Phymatotrichum omnivorum* Duggar

» **Rust**: *Uromyces decoratus* Syd.

Chapter - 38

Diseases of Mesta
Hibiscus sabdariffa var. *altissima* and *Hibiscus cannabinus*

S. No.	Disease	Pathogen
1.	Foot and stem rot	*Phytophthora parasitica* var. *sabdariffae*
2.	Stem rot	*Phyllosticta hibiscini*
3.	Yellow vein mosaic	*Mesta yellow vein mosaic virus*

1. Foot and stem rot

Symptoms

» The symptom appears on the stem generally a few inches above the ground but the spots may be seen at higher or lower level also.

» The spots are deep brown to blackish in colour with variable size.

» Larger spots very often girdle the stem and as a result the plant breaks at the point of infection.

» No fibre is obtained from such plants.

Chromista: *Phytophthora parasitica* var. *sabdariffae* (Waterhouse 1963)

Diseases of Field Crops and their Management

Favourable conditions

» It is favoured by high temperature (30-35°C)

» Continuous drizzling and water stagnation

» High rainfall and cloudy condition

Management

» Grow moderately resistant AMV 1, Roselle Type 1 and AP 481

» Red bristled *H. sabdariffa* lines are more resistant

» Seed treatment with mancozeb at 0.5 g/ kg

» Soil drenching with mancozeb 0.2%

» Spray copper oxychloride 50 WP at 5.0–7.0 g /lit or carbendazim 50 WP at 2.0 g/ lit.

2. Stem rot

Symptoms

» The disease appears as water soaked areas on any parts of stem which turn into brown patches that visible from distance.

» Initially the portion of the stem above or below the patches looks healthy.

» Finally the infection girdles the stem completely which extend as much as a foot or more.

» The rot causes the tissues to become soft and easily peel off into shreds.

» The portions above the affected part may ultimately wilted, diebacked and break away.

» The surface of the affected parts is covered with white stands of mycelia which form cushion in the axils of the branches.

» Black coloured sclerotia may be observed on this mycelial mat.

» The pith region may be filling with these hard scleroria.

» The sclerotia were also noticed in bolls.

Diseases of Mesta

Fungus: *Sclerotinia sclerotiarum* (Lib) de Bary.

The sclerotia survive in the soil and with falling of temperature in the month of December-January, it germinates and developed apothecia in which asci and ascospores are formed. Upon discharge the ascospore cause infection.

Manangement

> Soil drenching with copper oxychloride 50 WP 0.25%.

3. Leaf blight

Symptoms

> The disease starts as discoloured water soaked area mostly from the margin of leaf which increased towards inward direction and infect the petiole, through which it moves towards the stem.

> The infected leaf and petiole started yellowing and finally fall off.

> The infection spread very fast under high humid and rainfall condition and plants become defoliated.

> Under dry condition the infection become restricted and dark / black coloured dot like pycnidia developed on the leaf

Fungus: *Phyllosticta hibiscini*

The pathogen produce white coloured colony with distinguished ring like zones on which black coloured dot like pycnidia.

Favourable conditions

> High humidity (75-95% RH), high temperature (30-37°C) and high rainfall.

Manangement

> Spray chlorothalanil at 0.25%.

4. Yellow vein mosaic

Symptoms

» The characteristic symptom of the disease is yellowing of veins and veinlets followed by complete chlorosis of leaves.

» The flowers and fruits are malformed causing low seed yield.

Virus: *Mesta yellow vein mosaic virus* (MeYVMV)

Vector: Whitefly (*Bemisia tabaci*)

Manangement

» Spray Imidaclorprid 17.8 SL at 0.3% and Thiamethoxam 25 WG at 0.25%.

Minor diseases

» **Anthracnose** : *Colletotrichum hibisci* Poll.

» **Tip rot** : *Phoma* spp.

» **Root rot** : *Rhizoctonia bataticola* (Taub.) Butler

» **Eye rot of stem** : *Myrothecium roridum* Tode ex Fr.

» **Rust** : *Aecidium garkeanum* Henn.

» **Leaf spot** : *Alternaria dianthi* F. Stevens & J.G. Hall; *Cercospora abelmoschi* S. Narayan, Kharwar, R.K. Singh & Bhartiya; *Cercospora hibiscina* Ellis & Everhart

» **Powdery mildew** : *Leveillula taurica*

Chapter - 39

Diseases of Agave - *Agave sisalana* Perr.

S. No.	Disease	Pathogen
1.	Anthracnose	*Glomerella cingulata*
2.	Leaf spot	*Coniothyrium concentricum*
3.	Leaf blight	*Alternaria alternata*

1. Anthracnose

Symptoms

- » The symptoms are circular depressed dark colored spots with raised ring.
- » The spots may spread to the entire leaf.

Fungus: *Glomerella cingulata* (Stoneman) Spaulding & von Schrenk

Management

- » Remove and destroy the infected plants.
- » Spray copper oxychloride 0.25%.

2. Leaf spot

Symptoms

> » The symptoms are zoned light grayish brown spots.

> » The spots may form concentric rings with small fruiting bodies.

> » Finally, leaves may rot.

Fungus: *Coniothyrium concentricum* (Desm.)

Management

> » Remove and destroy the infected plants.

> » Spray copper oxychloride 0.25% or metalaxyl + mancozeb 0.2%.

3. Leaf blight

Symptoms

> » Disease symptoms on the leaves were appeared as round and golden brown spots.

> » As the disease progressed, spots coalesced, enlarged and developed towards the tip of the leaf as grayish coloured longitudinal streaks.

> » Tissue necrosis within these spots was observed which ultimately leads to drying and death of leaves.

Fungi: *Alternaria alternata* (Fr.) Keiss.; *Botrytis cinerea* Pers.; *Stagonospora gigantea* Heald & F.A. Wolf.

Management

> » Remove and destroy the infected plants.

> » Spray metalaxyl + mancozeb 0.2%.

Chapter - 40

Diseases of Ramie - *Boehmeria nivea* L.

S. No.	Disease	Pathogen
1.	*Cercospora* leaf spot	*Cercospora boehremia*
2.	Damping-off	*Rhizoctonia solani*
3.	Eye rot	*Myrothecium roridum*
4.	Cane rot	*Rhizoctonia bataticola*
5.	Anthracnose	*Colletotrichum gloeosporioides*
6.	Stem rot	*Rhizoctonia bataticola*

1. *Cercospora* leaf spot

Symptoms

- » The disease appears as circular to angular spot on the upper surface of the leaf.

- » Sometimes the spots which are 1-8 mm in diameter and dark brown to nearly black in colour are limited by the leaf veins.

- » The centre of older spots, however, turns paler and becomes greyish brown.

- » Infection is usually more severe on the lower leaves.

- » Leaves severely spotted turn yellow and fall prematurely.

340   Diseases of Field Crops and their Management

Fungus: *Cercospora boehremia*

Favourables conditions

» The development of this disease is favoured by moist and cool weather.

Management

» Spray copper oxychloride 0.2% or propiconazole 0.1% or difenoconazole 0.1%.

2.   Damping-off

Symptoms

» In seed beds (when ramie is raised through seed for breeding purpose) the seedlings may be attacked by damping off in humid and moist conditions.

» Affected seedlings are pale green and show a girdle of brown decaying cortex leading to collapsing of seedling.

Fungus: *Rhizoctonia solani* Kuhn.

Favourables conditions

» High soil moisture and relative humidity in soil.

Management

» Sterilization of soil of seed bed by using formaldehyde (1:50).

» Treating planting stalks with captan 0.2%.

3.   Eye rot

Symptoms

» The first symptom appears as irregular, small, round tan coloured spots about 1 mm in diameter on the upper surface of the lamina.

» The spots become circular and elongated to irregular, 1-16 mm in diameter and brown to dark brown in colour.

Diseases of Ramie

Fungus: *Myrothecium roridum* Tode ex Fr.

Management

» Spray copper oxychloride 4-5 g/ lit.

4. Cane rot

Symptoms

» The leaves then crinkle, rot, adhere to the canes and ultimately shed off.

» Root system become weak and turns brown.

» Brown shunken circular or elongated lesions are common on the stalks in the basal regions.

» They increase in size and several such lesions coalesce and girdle the stalk.

» When the lesions streaks along the length of the stalk it shrivels resulting in complete drying.

Fungus: *Rhizoctonia bataticola* (Taub.) Butler

Favourable conditions

» The disease is prevalent during rainy season on mature clumps.

Management

» Soil drenching with copper oxychloride 0.25%.

5. Anthracnose

Symptoms

» Lesions were initially small, scattered, round, and gray with brown margin on leaves.

» Irregular spots developed and expanded until the leaves withered.

Fungus: *Colletotrichum gloeosporioides* (Penz.) Penz. & Sacc.

Management

» Spray with carbendazim 0.2% or propiconazole 0.1% or difenoconazole 0.1%.

6.   Stem rot

Symptoms

» Symptoms of the disease were wilting and water soaking of the basal portion of the plant.

» Severely infected plant turned brown, defoliated and ultimately died.

» Profuse white mycelia of the fungus covered the infected stem.

Fungus: *Rhizoctonia bataticola* (Taub.) Butler.

Favourable conditions

» The disease is prevalent during rainy season on mature clumps.

Management

» Soil drenching with copper oxychloride 0.25%.

Chapter - 41

Diseases of Sugarcane
- *Saccharum officinarum* L.

S. No.	Disease	Pathogen
1.	Red rot	*Colletotrichum falcatum*
2.	Whip smut	*Ustilago scitaminea*
3.	Sett rot or pineapple disease	*Ceratocystis paradoxa*
4.	Wilt	*Cephalosporium sacchari*
5.	Rust	*Puccinia erianthi*
6.	Gummosis	*Xanthomonas axonopodis* pv. *vasculorum*
7.	Red stripe and top rot	*Acidovorax avenae* subsp. *avenae*
8.	Ratoon stunting	*Leifsonia xyli* subsp. *xyli*
9.	Grassy shoot	Candidatus *Phytoplasma oryzae*
10.	Sugarcane mosaic disease	*Sugarcane mosaic potyvirus*

1. Red rot

Symptoms

» Red rot is often referred as cancer of sugarcane

» The first external symptom appears mostly on third or fourth leaf which withers away at the tips along the margins.

344   Diseases of Field Crops and their Management

» Typical symptoms of red rot are observed in the internodes of a stalk by splitting it longitudinally.

» These include the reddening of the internal tissues which are usually elongated at right angles to the long axis of the stalk.

» The presence of cross-wise white patches are the important diagnostic character of the disease.

» The diseased cane also emits acidic-sour smell.

» As the disease advances, the stalk becomes hollow and covered with white mycelial growth.

» Later the rind shrinks longitudinally with minute black, velvetty fruiting bodies protruding out of it.

» The pathogen also produces tiny reddish lesions on the upper surface of leaves with dark dots in the centre.

» The lesions are initially blood red with dark margins and later on with straw coloured centres.

» Often the infected leaves may break at the lesions and hang down, with large number of minute black dots.

Fungus: Anamorph stage: *Colletotrichum falcatum* Went 1893
Teleomorph stage: *Physalospora tucumanensis*
Glomerella tucumanensis (Speg.) Arx & E. Mull. (1954)

The fungus produces thin, hyaline, septate, profusely branched hyphae containing oil droplets. The fungus produces black, minute velvetty acervuli with long, rigid bristle-like, septate setae. Conidiophores are closely packed inside the acervulus, which are short, hyaline and single celled. The conidia are single celled, hyaline, falcate, granular and guttulate. Fungus also produces large number of globose and dark brown to black perithecia with a papillate ostiole.

Asci are clavate, unitunicate and eight-spored. Large number of hyaline, septate, filiform paraphyses is also present among asci. Ascospores are ellipsoid or fusoid, hyaline, straight or slightly curved and unicellular which measure 18-22 μm $\times$ 7-8 μm.

Diseases of Sugarcane

Disease cycle

The fungus is sett-borne and also persists in the soil on the diseased clumps and stubbles as chlamydospores and dormant mycelium. The primary infection is mainly from infected setts. Secondary spread in the field is through irrigation water and cultivation tools. The rain splash, air currents and dew drops also help in the spread of conidia from the diseased to healthy plants in the field. The fungus also survives on collateral hosts *Sorghum vulgare, S. halepense* and *Saccharum spontaneum.* If the conidia settle on the leaves they may germinate and invade the leaves through various types of wounds. Stem infection may take place through insect bores and root primordia. The soil-borne fungus may also enter the healthy setts through cut-ends, and cause early infection of the shoots. Though the teleomorph stage of the fungus has been observed in nature, the role of ascospores in the Disease cycle is not understood.

Favourable conditions

- » Monoculturing of sugarcane.

- » Successive ratoon cropping.

- » Water logged conditions and injuries caused by insects.

- » Drought conditions during the initial growth phase.

- » High atmospheric humidity (90%) and pH 5-6.

- » Mean temperature range of 29.4 to 31°C is optimum for the development of the disease.

- » Continuous cultivation of same variety in the field

Management

- » Adopt crop rotation by including rice and green manure crops.

- » Select the setts from the disease free fields or disease free areas.

- » Avoid ratooning of the diseased crop.

- » Furrow application of fully grown bioagent in decomposed pressmud at 8 tonnes/ha.

- » Grow resistant varieties Co 0118 (Karan 2), Co 0232, Co 0237 (Karan 8), Co 0238 (Karan 4), Co 0239, Co 0124, Co 0704, Co 1148, Co 1214, Co 1158, Co 06027, Co 06030, Co 89003, Co 98014, Co 86249, Co 86010, Co 85019, CoC 99063, Co

62198, Co 7704, CoSi 94077, CoC 22, CoC 23, CoG 5, CoK 30, CoSi 6, Cos 443, CoS 8465, COL 9, Bo 17, Bo 141, Bo 110, Bo 154, Bo 153, CoP 9301, CoP 11437, Co 86249 (Bhavani), Co 2001-13 (Sulabh), Co 99004 (Damodar), Co 0403, Kalong, Dhansiri and moderately resistant varieties Co 8001 and Co 8201.

» Setts can be treated with aerated steam therapy (AST) at 52°C for 4 to 5 hours and by moist hot air at 54°C for 2 hours or treating the setts at 52°C or the soaking of setts in cold running water for 48 h followed by hot-water treatment (50°C for 150-180 min).

» Setts can be treated with moist hot air therapy (MHAT) of seed cane at 54°C for 4 hr (R.H. 95-100).

» Bleaching powder was effective in reducing red rot incidence

» Soak the setts in carbendazim 0.1% or triademefon 0.05% or thiophanate methyl 0.25% for 15 minutes before planting.

» Spray chlorothalanil 2 g/ lit or copper oxychloride 5 g/ lit or carbendazim 12% + mancozeb 36% at 1g/ lit.

2. Whip smut

Symptoms

» It is a culmiculous smut, causes total crop loss.

» The affected plants are stunted and the central shoot is converted into a long whip-like, dusty black structure.

» The length of the whip varies from few inches to several feet.

» In early stages, this structure is covered by a thin, white papery membrane.

» The whip may be straight or slightly curved.

» On maturity it ruptures and millions of tiny black smut spores (teliospores) are liberated and disseminated by the wind.

» Affected plants are usually thin, stiff and remain at acute angle.

» The whip like structure, representing the central shoot with its various leaves, may be produced by each one of the shoots/ tillers arising from the clump.

» The smutted clumps also produce mummified arrows in which lower portion consisted

Diseases of Sugarcane 347

of a normal inflorescence with typical flowers and the upper portion of the rachis is converted into a typical smut whip.

» Occasionally smut sori may develop on the leaves and stem.

Fungus: *Ustilago scitaminea* Syd., *Sporisorium scitamineum* (Syd.) M. Piepenbr.

The fungal hyphae are primarily intercellular and collect as a dense mass between the vascular bundles of host cell and produce tiny black spores. The thin membrane which covers the smut whip represents the host epidermis. The smut spores are light brown in colour, spherical, echinulated and measuring 6.5-8.5μm in diameter. Smut spores germinate to produce 3-4 celled, hyaline promycelium and produce 3-4 sporidia which are hyaline and oval shaped with pointed ends.

Disease cycle

Teliospores may survive in the soil for long periods, upto 10 years. The spores and sporidia are also present in the infected plant materials in the soil. The smut spores and dormant mycelium also present in or on the infected setts. The primary spread of the disease is through diseased seed-pieces (setts).

In addition, sporidia and spores present in the soil also spread through rain and irrigation water and cause soil-borne infection. The secondary spread in the field is mainly through the smut spores developed in the whips, aided by air currents. The fungus also survives on collateral hosts like *Saccharum spontaneum, S. robustum, Sorghum vulgare, Imperata arundinacea* and *Cyperus dilatatus.*

Favourable conditions

» Monoculturing of sugarcane.

» Continuous ratooning and dry weather during tillering stage.

Management

» Plant healthy setts taken from disease free area.

» Remove and destory the smutted clump (collect the whips in a thick cloth bag/ polythene bag and immerse in boiling water for 1 hr to kill the spores).

» Discourage ratooning of the diseased crops having more than 10 % infection.

348 Diseases of Field Crops and their Management

» Follow crop rotation with green manure crops or dry fallowing.

» Grow redgram as a companion crop between 2 rows of sugarcane.

» Grow resistant varieties like Co 312, Co 497, Co 449, Co 527, Co 997, Co 1008, Co 1132, Co 6928, Co 1111, Co 1118, Co 1148, Co 6806, Co 7704, Co 8338, Co 8341, Co 8019, CoB 94164, Co J 64, Bo 11, Bo 17, Bo 22, Bo 24, Bo 91, Bo 99, Co 86032 (Nayana), Co 86249 (Bhavani), Co 2001-15, Co 94012 and moderately resistant varieties CoC 85061 and CoC 8201.

» Grow red rot and smut resistant varieties like Co 06030, Co 0403, Co 2001-13 (Sulabh).

» To prevent the diseases and also to improve germination, the seed setts are dipped into methyl ethyl mercuric chloride (agallol 0.5%) or aretan (0.25%) or tafasan (6%) before planting for 10 minutes.

3. Sett rot or pineapple disease

Symptoms

» The disease primarily affects the setts usually two to three weeks after planning.

» The fungus is soil-borne and enters through cut ends and proliferates rapidly in the parenchymatous tissues.

» The affected tissues first develop a reddish colour which turns to brownish black in the later stages.

» The severely affected setts show internodal cavities covered with the mycelium and abundant spores.

» A characteristic pineapple smell is associated with the rotting tissues.

» The setts may decay before the buds germinate or the shoots may die after reaching a height of about 6-12 inches.

» Infected shoots are stunted.

Fungus: *Ceratocystis paradoxa* (Dade) C. Moreau, (1952)
Syn: *Thielaviopsis paradoxa* (De Seynes) Höhn., (1904);
Ceratostomella paradoxa Dade, Trans. (1928)

Diseases of Sugarcane

The fungus produces both macroconidia and microconidia. Conidiophores are linear, thin walled with short cells at the base and a long terminal cell. The microconidia are hyaline when young but become almost black at maturity. They are thinwalled, cylindrical and produced endogenously in chains in the long cells of conidiophores and pushed out in succession.

Macroconidia are produced singly or in chains on short, lateral conidiophores. Macroconidia are spherical or elliptical or truncate or pyriform and are hyaline to olive green or black measuring 16-19×10-12 um.

The fungus also produces chlamydospores on short lateral hyphae in chains, which are oval, thick walled and brown in colour. The perithecia are flask shaped with a very long neck. The bulbous base of the perithecium is hyaline or pale yellow, 200-300μm in diameter and ornamented with irregularly shaped, knobbed appendages. The ostiole is covered by numerous pale-brown, erect tapering hyphae. Asci are clavate and measures 25×10μm and ascospores are single celled, hyaline, ellipsoid, more convex on one side, measures 7-10 × 2.5-4μm.

Disease cycle

The fungus survives as conidia and chlamydospores in the soil and in the infected, burried cane tissues. The inoculum moves from field to field through wind-borne conidia or irrigation or rain water. Inside the sett it spreads rapidly through the parenchymatous tissues and causes sett rot.

The insects like cane borer (*Diatraea dyari*) also helps in the spread of the disease. The pathogen also survives on coconut, cocoa, mango, papaya, coffee, maize and arecanut. Insects also play a part in the dissemination of the pathogen.

Favourable conditions

> » Poorly drained fields.

> » Heavy clay soils

> » Temperature of 25-30°C

> » Prolonged rainfall after planting.

Management

» Soak the setts in carbendazim 0.05% for 15 min.

» Use long setts having 3 or 4 buds.

» Provide adequate drainage during rainy seasons.

4.   Wilt

Symptoms

» The first symptom of the disease is visible in the canes of 4-5 month age.

» The canes may wither in groups.

» The affected plants are stunted with yellowing and withering of crown leaves.

» The midribs of all leaves in a crown generally turn yellow, while the leaf lamina may remain green.

» The leaves dry up and stem develop hollowness in the core.

» The core shows the reddish discolouration with longitudinal red streaks passing from one internode to another.

» In severe cases, spindle shaped cavities tapering towards the nodes develop in each internode.

» The canes emit a disagreeable odour, with lot of mycelial threads of the fungus cover the cavity.

Fungi: *Cephalosporium sacchari* E.J. Butler & Hafiz Khan, (1913)
Fusarium sacchari (E.J. Butler & Hafiz Khan) W. Gams, (1971)

The fungal mycelium is hyaline, septate and thin walled. The conidiophores are simple or branched and produce single celled, hyaline, oval to elliptical microconidia.

Disease cycle

The fungus is soil-borne and remains in the soil as saprophyte for 2-3 years. The disease is primarily transmitted through infected seed pieces. The secondary spread is aided by wind, rain and irrigation water.

Diseases of Sugarcane 351

Favourable conditions

» High day temperature (30-35°C) and low humidity (50-60%).

» Low soil moisture and alkaline soils.

» Excess doses of nitrogenous fertilizers.

Management

» Select the seed material from the disease-free plots.

» Avoid the practice of ratooning in diseased fields.

» Burn the trashes and stubbles in the field.

» Grow red rot and wilt resistant varieties like Co 0118 (Karan 2), Co 0232 and Co 0233.

» Grow wilt resistant varieties like Co 8019, Co 8341, Co 8338 and Co 8019.

» Grow coriander or mustard as a companion crop in the l early stages of crop.

» Dip the setts in boran or manganese 40ppm for 10 minutes or in methoxyethyl mercuric chloride (emisan 0.25%) or carbendazim 0.05% for 15 minutes.

5.   Rust

Symptoms

» Minute, elongated, yellow spots (uredia), usually 2-10 × 1-3 mm appear on both the surfaces of young leaves.

» The pustules turn to brown on maturity.

» Late in the season, dark brown to black telia appear on the lower surface of leaves.

» In severse cases, the uredia also appear on the leaf sheath and the entire foliage looks brownish from a distance.

Fungi: *Puccinia erianthi* Padwick & A. Khan, (1944)
      Syn: *P. melanocephala* and *P. kuehnii*

The mycelium is hyaline, branched and septate. *P. kuehnii* produces ovoid or pear shaped, single celled uredospores measuring 29-57 × 8-37µm with apical thickening and golden yellow in colour. Teliospores are produced in scanty which are yellow in colour, club shaped, two celled, smooth walled and measuring 24- 34 × 18-25µm single celled, dark yellow coloured with 4 equatorial pores.

Teliospores are produced in abundance, which are pale to brick colour, two celled, smooth walled and slightly constricted at septum. Occurrence of pycnial and aecial stages and the role of alternate host are unknown.

Disease cycle

The fungus survives on collateral hosts like *Erianthus fulvus* and *Saccharum spontaneum.* The uredospores also survive in the infected stubbles in the soil. The disease is mainly spread through air-borne uredospores.

Favourable conditions

- » Temperature of 30°C and humidity between 70 and 90%.

- » High wind velocity and continuous cloudiness.

Management

- » Remove the collateral hosts.

- » Grow resistant varities like Co 86249 (Bhavani), Co 421, Co 467, Co 575, Co 603, Co 678, Co 732, Co 826 and Co 928.

- » Spray *Darulaca filum* 20 ml/ lit.

- » Spray oxycarboxin 1 kg or mancozeb 2 kg/ha.

6.   Gummosis

Symptoms

- » The bacterium produces two distinct types of symptoms.

- » On the mature leaves, longitudinal stripes or streaks, 3-7mm in width and several cm in length, appear around the affected veins, near the tip.

- » Initially these stripes are pale yellow in colour, later turn to brown.

Diseases of Sugarcane 353

> The affected tissues slowly dry up.

> The infected canes are stunted with short internodes, giving a bushy appearance.

> When such canes are cut transversely or split open longitudinally, dull yellow bacterial ooze comes out from the cut ends and bacterial pockets are seen inside the slitted cane.

> The fibro vascular bundles are deep red and internodal cavities formed in the severe cases are filled with yellow coloured bacterial gums.

Bacterium: *Xanthomonas axonopodis* pv. *vasculorum*

The bacterium is a short rod, Gram negative, non spore forming measuring 1.0-1.5μm × 0.4-0.5μm, with a single polar flagellum. It is facultative anaerobe and it produces yellow slimy growth.

Disease cycle

The bacterium remains viable in the soil as well as in infected canes. The primary transmission is through naturally affected diseased setts or through soil-borne contamination. The secondary spread may be through wind splashed rain, harvesting implements, animals and insects. The bacterium can survive in the insect's body for a long time and in this way may be transmitted long distances. On entry into the host the bacterium reaches the vascular tissues and becomes systemic. The bacterium also perpetuates on maize, sorghum, pearlmillet and other weed hosts, which also serve as sources of inoculum.

Management

> Remove and burn the affected clumps and the stubbles in the field.

> Select setts from disease free areas.

> Avoid growing collateral hosts like maize, sorghum and pearl millet near the sugarcane fields.

7. Red stripe and top rot

The disease is common in all sugarcane growing areas of world. In India it was first reported in 1933. In is more severe in Bihar, Uttar Pradesh, Punjab, Maharashtra and Tamil Nadu.

Symptoms

» The disease first makes it appearance on the basal part of the young leaves.

» The stripes appear as water soaked, long, narrow chlorotic streaks and become reddish brown in few days.

» These stripes are 0.5 to 1 mm in width and 5-100 mm in length, run parallel to the midrib.

» The stripes remain confined to lower half of the leaf lamina and whitish flakes spreads to growing points of the shoot and yellowish stripes develop, which later turn reddish brown.

» The rotting may commence from the tip of the shoot and spreads downwards.

» The core is discoloured to reddish brown and shrivelled and form cavity in the centre.

» In badly affected fields, a foul and nauseating smell appears.

Bacterium: *Acidovorax avenae* subsp. *avenae* (Manns) Willems *et al*. 1992
Pseudomonas rubrilineans Lee at al.

The bacterium is a short rod (0.7×1.67µm), gram negative, non capsulate with a polar flagellum.

Disease cycle

The pathogen remains viable in the soil and infected plant residues. The bacterium also survives on sorghum, pearlmillet, maize, fingermillet and other species of *Saccharum*. The bacterium primarily spreads through infected canes. The secondary spread is mainly through rainsplash, irrigation water and insects. Infected parenchymatous cells may collapse and normal functioning of the plant parts may fail. Several grasses, including ragi and bajra, have been reported to be infected by the bacteria and these hosts may also play a role in the perpetuation and spread of the pathogen.

Favourable conditions

» Continuous ratooning and prolonged rainy weather with low temperature (25°C).

Management

» Select setts from healthy fields and the affected plants should be removed and burnt.

Diseases of Sugarcane

» Growing resistant varieties Co 6805, Co 7202, Co 7321, Co 7537, Co 7642 and Co 8005.

» Avoid growing collateral hosts near the sugarcane fields.

8. Grassy shoot (Albino)

Symptoms

» The disease appears nearly two months after planting.

» The disease is characterised by the production of numerous lanky tillers from the base of the affected shoots.

» Leaves become pale yellow to completely chlorotic, thin and narrow.

» The plants appear bushy and 'grass-like' due to reduction in the length of internodes premature and continuous tillering.

» The affected clumps are stunted with premature proliferation of auxillary buds.

» Cane formation rarely occurs in the affected clumps, if formed, thin with shorter internodes having aerial roots at the lower nodes.

» The buds on such canes usually papery and abnormally elongated.

Bacterium: Candidatus *Phytoplasma oryzae*

The disease is caused by a phytoplasma. Two types of bodies are seen in ultrathin sections of phloem cells of infected plants. The spherical bodies are 300-400 nm diameter and filamentous bodies of 30-53 nm diameters in size.

Disease cycle

The primary spread of the phytoplasma is through diseased setts and cutting knifes. The pathogen is transmitted secondarily by aphid's *viz., Rhopalosiphum maydis, Melanaphis sacchari* and *M. idiosacchari.* Sorghum and maize serves as natural collateral hosts.

Management

» Avoid selection of setts from diseased area.

» Grow resistant varieties like Co 617.

356   Diseases of Field Crops and their Management

» Eradication of diseased parts as soon as symptoms are seen.

» Pre-treating the healthy setts with hot water at 52°C for 1 hour before planting

» Treating the healthy setts with hot air at 54°C for 8 hours.

» Spraying the crop twice a month with insecticides like malathion or triazophos 2ml/ lit.

9. Ratoon stunting

Symptoms

» Diseased clumps usually display stunted growth, reduced tillering, thin stalks with shortened internodes and yellowish foliage.

» Orange-red vascular bundles in shades of yellow at the nodes are seen in the infected canes.

Bacterium: *Leifsonia xyli* sub sp. *xyli* (Davis, Gillaspie, Vidaver & Harris) (Xylem inhabiting fastidious bacteria)

The pathogen (*Clavibacter xyli* sub sp. *xyli*) is known to be present in the xylem cells of infected plants. They are small, thin, rod shaped or coryneform (0.15 to 0.32μm wide and 1.0-2.7μm long) and Gram positive.

Disease cycle

The primary spread is through the use of diseased setts. The disease also spreads through harvesting implements contaminated with the juice of the diseased canes. Maize, sorghum, Sudan grass and *Cynodon* serve as collateral hosts for the pathogen.

Management

» Select the setts from disease free fields or from disease free commercial nursery.

» Remove and burn the clumps showing the disease incidence.

» Treat the setts with hot water at 50°C for 2 hrs or 52°C for 30 mins.

» Treat the setts with hot air at 54°C for 8 hrs

Diseases of Sugarcane

10. Sugarcane mosaic disease

Symptoms

» The disease appears more prominently on the basal portion of the younger foliage as chlorotic or yellowish stripes alternate with normal green portion of the leaf.

» As infection becomes severe, yellow stripes appear on the leaf sheath and stalks.

» Elongated necrotic lesions are produced on the stalks and stem splitting occurs.

» The necrotic lesions also develop on the internodes and the entire plant becomes stunted and chlorotic.

Virus: *Sugarcane mosaic virus* (*Potyvirus*)

Sugarcane mosaic potyvirus is a flexous rod, 650-770nm long × 12-15nm with ss RNA genome.

Disease cycle

The virus is mainly transmitted through infected canes used as seed. The virus also infects *Zea mays* and a number of other cereals (*Sorghum vulgare, Pennisetum americanum, Eleusine indica, Setaria lutescens, Echinochloa crusgalli, Stenotaphrum secondatum, Digitaria didactyla*) which serve as potential sources of virus inoculum. The virus also spreads through viruliferous aphids *viz., Melanaphis sacchari, Rhopalosiphum maidis* in a non-persistant manner. The virus is also sap-transmissible. The incubation period varies from 7 to 20 days, depending upon the host variety and virus strain. The symptoms may be prominent or masked depending on the environmental conditions and variety.

Management

» Roguing of infected plants and use of disease free planting material.

» Grow mosaic-resistant Co 0237 (Karan 8) or, at least, tolerant varieties.

» Chemical sprays to manage the insect vector population in early crop stage.

» *Saccharum spontaneum* L. and *S. barberi* (Jesweit) carry resistance to mosaic.

» Rogue out the diseased clumps periodically. Select setts from the healthy fields as the virus is sett-borne.

Minor diseases

Damping-off: *Pythium aphanidermatum (Edson) Fitzp, P. debaryanum* R. Hesse,, *P. graminicola* Subraman., *P.ultimum* Trow.

- » Germinating seeds and young seedlings are attacked and killed.
- » In pre-emergence phase and seedlings show water soaked lesions at collar region, leading to withering and drying in post emergence stage.

Downy mildew: *Peronosclerospora sacchari* (T. Miyake) Shirai & Hara

- » Downy fungal growth with yellow stripes on upper surface, shredding of older leaves, rapid elongation of internodes of affected canes.

Eye spot: *Helminthosporium sacchari* E.J. Butler

- » The water soaked spot develops on leaves, later elongated and turns to form "eye" shaped spot with reddish brown centre surrounded by straw yellow tissues.

Ring spot: *Leptosphaeria sacchari* Speg.

- » The water soaked spots appear on leaves and turns to straw colour later surrounded by a thin reddish brown band and a diffused discolouration zone.

Leaf scald: *Xanthomonas albilineans* (Ashby) Dowson

- » Whitish lines appear on the leaves, run to the full length of leaves and sheaths.
- » Later leaves wither and dry from tip down-wards, give a scald appearance to the clump.
- » Sprouting of lateral buds of the matured canes occurs in acropetal fashion.

White leaf: Candidatus *Phytoplasma*

- » Sugarcane white leaf is of minor importance and is caused by phytoplasma.
- » The plants exhibit pure white leaves, stripped leaves and mottled leaves.
- » Its vector is *Matsumuratettix hiroglyphicus.*

Diseases of Sugarcane

Phanerogamic parasite: *Striga euphrasioides* (Vahl) Benth.

» Partial root parasite, growing up from the roots to form a leafy shoots.

» The parasite synthesis carbohydrates through the green chlorophyll pigments in the leaves but its other nutrients it depends on host root.

» It is usually controlled by pulling out the shoots before flowering and seed set.

Chapter - 42

Diseases of Sugarbeet - *Beta vulgaris* L.

S. No.	Disease	Pathogen
1.	*Aphanomyces* seedling disease	*Aphanomyces cochlioides*
2.	Damping off and Root rot	*Pythium aphanidermatum*
3.	*Rhizoctonia* Damping off	*Rhizoctonia solani*
4.	*Rhizoctonia* root rot	*Rhizoctonia bataticola*
5.	*Phoma* disease	*Phoma betae*
6.	*Cercospora* leaf spot	*Cercospora beticola*
7.	*Sclerotium* root rot	*Sclerotium rolfsii*
8.	*Alternaria* leaf spot	*Alternaria tenuis*
9.	Bacterial scab	*Streptomyces scabies*
10.	Bacterial leaf spot	*Pseudomonas syringae*

1. *Aphanomyces* seedling disease

Symptoms

» Generally, seedling emergence is not affected, but 1–3 weeks after emergence a dark grey, watersoaked lesion develops on the hypocotyl.

Diseases of Sugarbeet

» The lesion expands rapidly, and soon the entire hypocotyl appears dark grey or brown to black and shrinks to become threadlike, hence the term 'blackleg' or 'black root'.

» Infected seedlings are stunted and have reduced vigour; they may fall over and die or be broken off by the action of wind, but often they survive and show some recovery.

» Plants surviving disease at the seedling stage may develop the chronic root rot phase

Chromista: *Aphanomyces cochlioides* Drechs.

Hyphae are hyaline, 3-9 µm in diameter, and coenocytic. Slender, irregular, filamentous sporangia (upto 3-4 mm in length) are produced at right angles to the parent hypha. Primary zoospores differentiate within sporangia, are extruded and encyst in clusters at the ends of long evacuation tubes from sporangial elements. Biflagellate, reniform, secondary zoospores emerge from primary zoospore cysts and, after a period of motility, these encyst and finally germinate by germ tube.

Disease cycle

Oospores survive for long periods in soil or infected plant debris. Under conditions of high soil moisture, oospores germinate by germ tube, which can directly infect the host, or produce an apical sporangium giving rise to zoospores. The pathogen can be spread in infested soil and locally by movement of asexual zoospores.

Favourable conditions

» High soil moisture and free water

» Soil temperatures increased from 18 to 32°C, with an optimum around 25° C.

Management

» High soil fertility, especially high levels of phosphorus, promotes rapid seedling growth and reduces the severity of black root.

» Rotation with non-susceptible crops such as maize, soybean, potatoes, or small grains.

» Soil drenching with metalayl + macnozeb 0.2%.

362   Diseases of Field Crops and their Management

2. *Pythium* Damping off and Root rot

Symptoms

Damping off: It causes pre-emergence and post- emergence damping off.

However, water-soaked, grey-black lesions can also develop on the hypocotyls of germinated seedlings, causing post-emergence damping-off.

Root rot: This phase is characterized by yellowing, wilting and drying of the plants involving all the aerial parts. Infected roots show externally a deep brown discolouration.

Chromista: *Pythium aphanidermatum* (Edson) Fitzp.
In addition to *P. aphanidermatum, P. debaryanum* and *P. ultimum* are also responsible for causing damping off and root rot in tropics.

P. sylvaticum Campbell and Hendrix and *P. intermedium* de Bary in temperate.

Pythium spp. are ubiquitous in agricultural soils and can cause plant losses wherever high soil moisture and other factors impede seed germination and emergence.

Management

> Seed treatment with metalaxyl, thiram or hymexazol is very effective.

3. *Rhizoctonia* Damping off

Symptoms

> The fungus can induce some pre-emergence damping-off but usually affects seedlings after emergence.

> A dark brown lesion begins just below the soil surface and extends up the hypocotyl, with a sharp line between diseased and healthy tissue.

> When the hypocotyl is girdled, the seedling collapses and dies.

Fungus: *Rhizoctonia solani* Kühn [Teleomorph, *Thanatephorus cucumeris* (Frank) Donk]

Hyphae are pale to dark brown, branching near the distal septum of hyphal cells, often nearly at right angles; branch hyphae are commonly constricted at the point of origin.

Diseases of Sugarbeet 363

Aggregates of thick-walled, dark brown monilioid cells ('barrel-shaped cells', 'bulbils') are also produced. Individual cells are multinucleate and have a prominent dolipore septal apparatus.

Favourable conditions

- » Sclerotia have been reported to survive in soil for several years

- » Warm soil

Management

- » Seed treatment with metalaxyl 0.2% or mancozeb 0.2%.

4. *Rhizoctonia* root rot

Symptoms

- » It causes crown rot and dry root rot.

- » In crown rot, basal portion of the petiole blackens and subsequently the entire crown of the plant rots.

- » In dry root rot, roots in upper 1-2 cm layer show sunken lesions.

- » Beneath these lesions pockets of spongy tissues develop.

- » *R. bataticola:* Numerous small, black, round sclerotia are seen on the diseased roots. On the petiole pycnidia are seen.

Fungus: *Rhizoctonia solani* – The sclerotia are irregular, brown to black and 5 mm in dia. *R. bataticola* – The sclerotia measure up to 1 mm in dia. In the pycnidial stage, the fungus produces globose pycnidia and hyaline, oval to elliptical pycnidiospores.

5. *Phoma* disease: *Phoma betae*

Seedling blight: Diseased seedlings exhibit black lesions on primary roots, just below the collar region leads to severe necrosis of root tip.

Leaf spot: Necrotic spots of circular to oval with light to dark brown concentric rings with diffused margin appear on the leaves.

Stem rot: Severe infection of basal stem leads to withering of plants.

Storage rot: Tubers from diseased plants carry incipient infection of the fungus and cause rotting during storage.

6. ***Cercospora* leaf spot:** *Cercospora beticola* Sacc.

Symptoms

- » Small, relatively circular lesions (2-3 mm), in the centre bright grey encircled with a reddish-brown margin.
- » As the disease progresses individual spots coalesce resulting in the death of large parts of the leaf blade and finally causing the wilting of numerous leaves.
- » Under a magnifying glass a grey mycelium with black dots (conidiophores) is visible.
- » Symptoms of the damage are first observed only on single plants, their spreading over the whole field is transmitted by rain-splash and wind.

7. ***Sclerotium* root rot:** *Sclerotium rolfsii* Sacc.

- » The disease attacks the plant and causes yellowing and wilting.
- » The fungus causes rotting of roots and tubers. Root rot affected plants can be easily pulled out.

8. ***Alternaria* leaf spot:** *Alternaria tenuis* Nees

- » Beginning at the edge or tip the leaves become brown in the intercostal fields and die.
- » A black, velvety coat develops on the tissue (bearer of conidia).

9. **Bacterial scab:** *Streptomyces scabies*

- » Small, brownish and slightly raised spots appear on fleshy tap root. Later they enlarge, coalesce and become very corky.
- » The rounded, wart-like growths are sparsely scattered over the tap root and often they are concentrated in bands.
- » Two types of scab *viz.*, shallow and deep scabs are found in sugerbeet.

Diseases of Sugarbeet

10. Bacterial leaf spot: *Pseudomonas syringae*

» Brown to black spots of irregular form and size mainly on the leaf edges or in hollows of the leaf blade.

» The dead tissue in the centre of the spots often becomes brittle.

» The damage incurred can recover under dry and warm conditions.

Minor diseases

Seedling blight: *Pythium, Aphanomyces, Phoma, Rhizoctonia*

Leaf blight: *Cercospora* sp., *Alternaria* sp.

Fusarium yellows: *Fusarium* sp.

   Beet mosaic virus

   Beet curly top virus

Chapter - 43

Diseases of Tobacco - *Nicotiana tabacum* L.

S. No.	Disease	Pathogen
1.	Damping-off	*Pythium aphanidermatum*
2.	Black shank	*Phytophthora parasitica* var. *nicotianae*
3.	Frog eye spot	*Cercospora nicotianae*
4.	Powdery mildew	*Erysiphe cichoracearum* var. *nicotianae*
5.	Brown spot	*Alternaria longipes*
6.	Anthracnose	*Colletotrichum tabacum*
7.	Wild fire	*Pseudomonas tabaci*
8.	Mosaic	*Tobacco mosaic virus*
9.	Leaf curl	*Tobacco leaf curl virus*
10.	Broom rape	*Orobanche cernua* var. *desertorum*

1. Damping-off

Symptoms

- » The pathogen attacks the seedlings at any stage in the nursery.

- » Sprouting seedlings are infected and wither before emergence from the soil (Pre emergence damping off).

Diseases of Tobacco 367

» Water soaked minute lesions appear on the stems near the soil surface, soon girdling the stem, spreading up and down in the stems and with in one or two days stem may rot leading to toppling over of the seedlings (Post-emergence damping off).

» The young seedlings in the nursery are killed in patches and infection spreads quickly.

» Under the favorable conditions, the entire seedlings in the nursery are killed within 3 to 4 days.

» A thick weft of mycelium may be seen on the surface of the soil.

Chromista: *Pythium aphanidermatum* (Edson) Fitzp., (1923)

The fungus produces thick, hyaline, thin walled, non-septate mycelium. It produces irregularly lobed sporangia which germinate to produce vesicle containing zoopores. The zoospores are kidney shaped and biflagellate. Oospores are spherical, light to deep yellow or yellowish brown coloured, measuring 17-19µm in diameter.

Disease cycle

The pathogen survives in the soil as oospores and chlamydospores. The primary infection is from the soil-borne fungal spores and secondary spread through sporangia and zoospores transmitted by wind and irrigation water.

Favourable conditions

» Over crowding of seedling.

» Ill drained nursery beds

» Heavy shade in nursery

» High atmospheric humidity (90-100%) and high soil moisture

» Low temperature (below 24°C) and low soil temperature of about 20°C.

Management

» Prepare raised seed beds with adequate drainage facility.

» Burn the seed beds with paddy husk before sowing.

368   Diseases of Field Crops and their Management

» Drench the seed bed with Bordeaux mixture 1% or copper oxychloride 0.2%, two days before sowing.

» Avoid over crowding of seedlings by using recommended seed rate (1-1.5 g/ 2.5m^2).

» Avoid excess watering of the seedlings.

» Spray the nursery beds two weeks after sowing with Bordeaux mixture 1% or copper oxychloride 0.2% or mancozeb 0.2% and repeat subsequently at 4 days interval under dry weather and at 2 days interval under wet cloudy weather or spray metalaxyl 0.2% at 10 days interval commencing from 20 days after germination.

2.   Black shank

Symptoms

» The pathogen may affect the crop at any stage of its growth.

» Even though all parts are affected, the disease infects chiefly the roots and base of the stem.

» Seedlings in the nursery show black discolor of the stem near the soil level and blackening of roots, leading the wet rot in humid condition and seedling blight in dry weather with withering and drying of tips.

» The pathogen also spreads to the leaves and causes blighting and drying of the bottom leaves.

» In the transplanted crop, the disease appears as minute black spot on the stem, spreads along the stem to produce irregular black patches and often girdling occurs.

» The upward movement leads to development of necrotic patches on the stems.

» The infected tissues shrink, leaving a depression and in advanced condition the stem shrivels and plant wilts.

» When the affected stem is split open, the pith region is found to be dried up in disc-like plates showing black discolouration.

» On the leaves large brown concentrically zonate patches appear during humid weather, leading to blackening and rotting of the leaves.

Chromista: *Phytophthora parasitica* var. *nicotianae* Breda de Haan, (1896)

Diseases of Tobacco

The fungus produces hyaline and non-septate mycelium. The sporangia, which are hyaline, thin walled, ovate or pyriform with papillae, develop on the sporangiophores in a sympodial fashion. Sporangia germinate to release zoospores which are usually kidney shaped biciliate and measure $11\text{-}13 \times 8\text{-}9\mu m$. The fungus also produces globoose and thick walled chlamydospores, measuring $27\text{-}42\mu m$ in diameter. Oospores are thick walled, globose, smooth and light yellow coloured, measuring $15\text{-}20\mu m$ in diameter.

Favourable conditions

» Frequent rainfall and high soil moisture.

» High population of rootknot nematodes *Meloidogyne incognita* var. *acrita.*

Disease cycle

The fungus lives as a saprophyte on organic wastes and infected crop residues in soil. The fungus is also present in the soil as dormant mycelium, oospores and chlamydospores for more than 2 years. The primary infection is by means of oospores and chlamydospores in the soil. Secondary spread is by wind-borne sporangia. The pathogen in the soil spreads through irrigation water, transport of soil, farm implements and animals.

Management

» Cover the seed beds with paddy husk or groundnut shell at 15-20 cm thick layer and burn.

» Provide adequate drainage in the nursery.

» Drench the nursery beds with Bordeaux mixture 1% or copper oxychloride 0.2%, two days before sowing.

» Spray the beds two weeks after sowing with metalaxyl 0.2% or captafol 0.2% or copper oxychloride 0.25% or Boreaux mixture 1% and repeat after 10 days.

» Select healthy, disease free seedlings for transplanting.

» Remove and destroy the affected plants in the field.

» Spray mancozeb 2 kg or copper oxychloride 1 kg or ziram 1 lit/ ha.

» Spot drench with Bordeaux mixture 0.4% or copper oxychloride 0.2%.

3. Frog eye spot

Symptoms

- » The disease appears mostly on mature, lower leaves as small ashy grey spots with brown border.

- » The typical spots have a white centre, surrounded in succession by grey, brown portions with a dark brown to black margin, resembling the eyes of a frog.

- » Under favorable conditions, several spots coalesce to form large necrotic areas, causing the leaf to dry up from the margin and wither prematurely.

- » Both yield and quality are reduced greatly.

- » The disease may occur in the seedlings also, leading to withering of leaves and death of the seedlings.

Fungus: *Cercospora nicotianae* Ellis & Everh., (1893)

The mycelium is intercellular and collects beneath the epidermis and clusters of conidiophores emerge through stomata. The conidiophores are septate, dark brown at the base and lighter towards the top bearing 2-3 conidia. The conidia are hyaline, slender, slightly curved, thinwalled and 2-12 septate.

Disease cycle

The pathogen is seed-borne and also persists on crop residues in the soil. The primary infection is from the seed and soil-borne inoculum. The secondary spread is through wind-borne conidia.

Favorable conditions

- » Temperature of 20-30°C.

- » High humidity (80-90%).

- » Close spacing, frequent irrigation and excess application of nitrogenous fertilizers.

Management

- » Remove and burn plant debris in the soil.

Diseases of Tobacco 371

- » Avoid excess nitrogenous fertilization.

- » Adopt optimum spacing.

- » Regulate irrigation frequency.

- » Spray the crop with Bordeaux mixture 0.4% or thiophanate methyl 750 g/ ha or carbendazim 750 g/ ha and repeat after 15 days.

4. Powdery mildew

Symptoms

- » Initially the disease appears as small, white isolated patches on the upper surface of the leaves.

- » Later, it spreads fast and covers the entire lamina.

- » The disease initially appears on the lower leaves and as disease advances, the rest of the leaves are also infected and sometimes powdery growth can be seen on the stem also.

- » The affected leaves turn to brown and wither and show scorched appearance.

- » The severe infection leads to defoliation and reduction in quantity and quality of the curable leaves.

Fungus: *Erysiphe cichoracearum* var. *nicotianae* (Comes) Jacz.
Syn: *Golovinomyces cichoracearum* var. *cichoracearum* (DC.) V.P. Heluta

The fungus is ecotophytic and produces hyaline, septate and highly branched mycelium. Short, stout and hyaline conidiophores arise from the mycelium and bear conidia in chains. The conidia are barrel shaped or cylindrical, hyaline and thin walled. Cleistothecia are black, spherical with no ostiole, with numerous densely-woven septate, brown-coloured appendages. They contain 10-15 asci which are ovate with a short stalk. Each ascus contains two ascospores which are oval to elliptical, thinwalled, hyaline and single celled.

Disease cycle

The fungus remains dormant as mycelium and cleistothecia in the infected plant debris in soil. The primary infection is mainly from soil-borne inoculum. The secondary spread is aided by wind blown conidia.

Favourable conditions

» Humid cloudy weather.

» Low temperature (16-23°C).

» Close planting and excess doses of nitrogenous fertilizers.

Management

» Apply balanced ferilizers.

» Avoid overcrowding of plants.

» Remove and destroy the affected leaves.

» Plant early in the season so that crop escapes the cool temperature at maturity phase.

» Spray carbendazim at 500 g/ ha.

5.   Brown spot

Symptoms

» Brown spot in contrast to frog-eye spot is not normally observed in the nursery but is very much prevalent in the field.

» Initially it appears on lower and older leaves as small brown, circular lesions, which spread, to upper leaves, petioles, stalks and capsules even.

» In warm weather (30°C) under high humidity, the leaf spots enlarge, 1-3 cm in diameter, centres are necroses and turn brown with characteristic marking giving target board appearance with a definite outline.

» In severe infection spots enlarge, coalesce and damage large areas making leaf dark-brown, ragged and worthless.

» On leaves nearing maturity, leaf spots are surrounded by bright yellow halo, due to production of toxin 'alternin' by the fungus.

Fungus: *Alternaria longipes* (Ellis & Everh.) E.W. Mason

Disease cycle

The fungus over summers in the soil as mycelium in the diseased plant debris such as stems

Diseases of Tobacco 373

of tobacco, weeds and other hosts. Under favourable weather in the next season conidial production starts which infect the lowermost leaves. As the season progresses, repeated infection cycles of the fungus attack healthy tissues of all aerial parts of tobacco of any age under high humidity. There is enormous spore density in the air near the end of the harvesting. Fungus persists as a mycelium in dead tissue for several months.

Management

» Removal and destruction of diseased plant debris.

» Continuous growing of tobacco must be avoided in the heavily infected fields.

» Spray maneb or zineb at 2 g/ ha or thiophanate methyl at 1 kg/ ha.

6. Anthracnose

Symptoms

» Initially, infection starts on lower leaves as pale-brown circular spots of 0.5 mm diameter with papery depressed centres outlined by slightly raised brown margin.

» The leaf-spots may remain small with white areas in the centre or coalesce to form large necrotic lesions.

» Under continuous humid weather, dark brown or black, elongated, sunken necrotic lesions appear on midrib, petiole and stem resulting in petiole and stem rot.

» Such seedlings do not establish in the field if planted.

» Primary infection starts from affected bits of aerial parts left in the soil in the previous season.

Fungus: *Colletotrichum tabacum* Böning (1932)

The pathogen is not seed-borne but persists in the soil on dried plant debris.

Management

» Raised seed beds and rabbing with farm wastes help in reducing the initial infection

» Removal and destruction of all diseased debris minimises the pathogen in the soil.

» Rogueing diseased seedlings especially with necrotic lesions on stem.

374 Diseases of Field Crops and their Management

» Protective spraying with Bordeaux mixture at 1.0% (2-2-500) or zineb at 2 kg/ha.

» Spraying of tricyclozole 2 ml/ lit.

7. Wild fire

Symptoms

» The leaf spots may occur at any stage of plant growth including the nursery seedlings.

» Dark brown to black spots with a yellow halo spreads quickly causing withering and drying of leaves.

» In advanced cases, lesions develop on the young stem tissues leading to withering and drying of the seedlings.

» In the fields, initially numerous water soaked black spots appear and latter become angular when restricted by the veins and veinlets.

» Several spots may coalesce to cause necrotic patches on the leaves.

» In advanced conditions, the entire leaf is fully covered with enlarged spots with yellow haloes.

» The leaves slowly wither and dry. Under humid weather condition, the disease spreads very fast and covers all the leaves and the entire plant gives a blighted appearance.

Bacterium: *Pseudomonas tabaci* (Wolf & Foster) Stevens
Syn: *Pseudomonas syringae pv. tabaci* (Wolf & Foster) Young, Dye & Wilkie

The bacterium is a rod, motile with a single polar flagellum, non-capsulated, non spore forming and Gram negative.

Disaease cycle

The bacterium survives in the infected crop residues in the soil, which is the primary source of infection. The secondary spread of the pathogen in the field is through wind splashed rain water and implements.

Favourable conditions

» Close planting.

Diseases of Tobacco

» Humid wet weather.

» Strong winds.

Management

» Remove and burn the infected crop residues in the soil.

» Avoid very close planting.

8. Mosaic

Symptoms

The disease begins as light discoloration along the veins of the youngest leaves. Soon the leaves develop a characteristic light and dark green pattern, the dark green areas associated more with the veins, turning into irregular blisters.

The early infected plants in the season are usually stunted with small, chlorotic, mottled and curled leaves. In severe infections, the leaves are narrowed, puckered, thin and malformed beyond recognition, Later, dark brown necrotic spots develop under hot weather and this symptom is called "Mosaic burn" or "Mosaic scorching".

Virus: *Tobacco mosaic virus* (TMV) *Tobamovirus*

It is a rigid rod measuring $300 \times 150\text{-}180$ nm with a central hollow tube of about 4nm diameter with ssRNA as its genome.

Disease cycle

The virus spreads most rapidly by contact wounds, sap and farm implements and operators. The virus remains viable in the plant debris in the soil as the source of inoculum as the longevity of the virus is very high. It is capable of remaining infective when stored dry for over 50 years. The virus has a wide host range, affecting nearly 50 plant species belonging to nine different families. The virus is not seed-transmitted in tobacco but tomato seeds transmit the virus. No insect vector known to transmit the virus.

» Tobacco mosaic is caused by *Nicotiana virus I (Marmor tabaci* var. *vulgare)*.

» It is a rod shaped particle measuring $300 \times 150\text{-}180\mu m$ with a central hollow tube of about 4 μm diameter.

- » It is made up of centrally placed Ribonucleic acid molecules (RNA) covered with a protein coat.
- » It is capable of remaining infective when stored dry for over 50 years.
- » The thermal inactivation point (TIP) of the virus is 90°C for 10 minutes.

Management

- » Remove and destroy infected plants.
- » Keep the field free of weeds which harbour the virus.
- » Wash hands with soap and running water before or after handling the plants or after weeding.
- » Grow resistant varieties like Jayasri, TMV RR2, TMV RR 2a and TMV RR3.
- » Prohibit smoking, chewing and snuffing during field operations.
- » Spray the nursery and main field with botanical leaf extracts of *Bougainvillea* or *Basella alba* at 1 litre of extract in 150 litres of water, two to three times at weekly intervals.
- » Adopt crop rotation by growing non-host plants for two seasons.

9. Leaf curl

Symptoms

- » The infections may occur at any stage, when young plants are infected the entire plant remains very much dwarfed.
- » Curling of leaves with clearing and thickening of veins, twisting of petioles, puckering of leaves, rugose and brittle and development of enations are the important symptoms of tobacco leaf curl disease.
- » Three forms of leaf curl expression are observed.
- » First the leaf margins curl downward towards the dorsal side and show thickening of veins with enation on the lower surface.
- » Second crinkle form shows curling of whole leaf edge towards dorsal side with enation

Diseases of Tobacco

on the veins and the lamina arching towards the ventral side between the veinlets.

» Third the transparent symptom shows the curling of leaves towards the ventral side with clearing of the veins and enations are absent.

Virus: *Tobacco leaf curl virus* (TLCV)

It is caused by Tobacco leaf curl geminivirus. Virions are geminate, non- enveloped, 18 nm diameter circular ssDNA genome.The virus is a white fly transmitted Geminivirus with ssDNA as genome.

» The virus is spherical measuring 35 μm in diameter.

» The virus is Nicotiana virus 10 or *Ruga tabaci*.

Disease cycle

The virus has a narrow host range in eight plant families. The virus is not transmissible through sap or seed. The whitefly, *Bemisia tabaci* is the vector. Due to wide host range of the virus many other plants are acting as source of inoculums.

Management

Remove and destroy the infected plants.
Rogue out the reservoir weed hosts which harbour the virus and whiteflies. Planting tobacco crop during the crop periods when the vector population is low.
Spray methyldemeton at 0.1 to 0.2 % to control the vectors.

10. Broom rape (Phanerogamic parasite)

Symptoms

» The affected tobacco plants are stunted and show withering and drooping of leaves to wilting.

» These indicate underground parasitism of the tobacco roots by the parasite.

» The young shoot of the parasite emerges from the soil at the base of the plants 5-6 weeks after transplanting.

» Normally, it appears on clusters of 50-100 shoots around the base of a single tobacco plant.

» The plants which are attacked very late exhibit no external symptoms but the quality and yield of leaves are reduced.

Parasite: *Orobanche cernua* var. *desertorum*

It is a total root parasite. It is an annual, fleshy flowering plant with a short, stout stem, 10-15 inches long. The stem is pale yellow or brownish red in colour and covered by small, thin, brown scaly leaves and the base of the stem is thickened. White-coloured flowers appear in the leaf axils. The floral parts are well developed with a lobed calyx, tubular corolla, superior ovary, numerous ovules and a large four-lobed stigma. The fruits are capsules containing small, black, reticulate and ovoid seeds.

Disease cycle

The seeds of the parasite remain dormant in the soil for several years. Primary infection occurs from the seeds in the soil. The seeds spread from field to field by irrigation water, animals, human beings and implements. The dormant seeds are stimulated to germinate by the root exudates of tobacco and attach it, to the roots by forming haustoria. Later, it grows rapidly to produce shoot and flowers. *Orobanche* also attacks other crops like brinjal, tomato, cauliflower, turnip and other cruciferous crops.

Management

» Rogue out the tender shoots of the parasite before flowering and seed set.

» Spray the soil with copper sulphate 0.25%

» Spray allyl alcohol 0.1% and apply few drops of kerosene directly on the shoot.

» Grow decoy or trap crops like chilli, moth bean, sorghum or cowpea to stimulate seed germination and kill the parasite.

Chapter - 44

Diseases of Jatropha - *Jatropha curcas* L.

S. No.	Disease	Pathogen
1.	Leaf spot	*Alternaria ricini*
2.	Rust	*Melampsora ricini*
3.	*Cercospora* leaf spot	*Cercospora jatrophicola*
4.	*Fusarium* wilt	*Fusarium oxysporum*
5.	*Botrytis* rot	*Botrytis ricini*
6.	Bacterial spot	*Xanthomonas ricinicola*

1.   **Leaf spot:** *Alternaria ricini* (Yoshii) Hansf.

»   Initially the attack appears on the leaves.

»   If the humidity is high, this disease develops very quickly on the fruit capsule and the fruit become black.

»   If the attack occurs in the beginning of flower forming, the buds may die.

»   And if it occurs at the end of flowering, the flower is opened but no fruit capsule is formed.

2. **Rust:** *Melampsora ricini, Phakopsora jatrophicola* (Arthur) Cummins

Syn: *Phakopsora arthuriana* Buriticá & J. F. Hennen)

» The disease looks like rust spots on the lower surface of the leaf and yellow in color

» A heavy damage may dry the leaf

3. *Cercospora* **leaf spot:** *Cercospora* sp.

» The damage caused by this disease on *Jatropha curcas* can be very heavy

» The symptom is that black spot or brown spot surrounded by ring with pale green in color

» The spots are found on both leaf surfaces

» When the spot is getting bigger, the spot is become gray, surrounded by brown color

» Initially the spot is round and finally the spot form is irregular

Fungi: *Cercospora jatrophicola* (Speg.) Chupp,
Cercospora jatrophigena U. Braun
Pseudocercospora jatrophae-curcas (J.M. Yen) Deighton
Pseudocercospora jatrophae (G.F. Atk.) A.K. Das & Chattopadh.
Pseudocercospora jatropharum (Speg.) U. Braun

4. *Fusarium* **wilt:** *Fusarium oxysporum* Schlecht. emend. Snyder & Hansen

» If the seedling is attacked, the leaves become pale green, and withered and finally died

» Leaves on the lower part of plant will fall and only leaves on the upper part will left.

» Complete death of the plant was observed.

5. *Botrytis* **rot:** *Botrytis ricini* N.F.Buchw. (1949)

» Initially the symptom is blackish spot on the flower

» It becomes a serious problem in the rainy season when the fruit capsule is formed

» The disease is well developed in a cool temperature and rain

Diseases of Jatropha

> The infected flower will be spoiled and covered by gray mold.

6. **Bacterial spot:** *Xanthomonas ricinicola* (Elliott) Dowson)

> This disease is caused by bacteria which may attack the cotyledon and the leaf

> The symptom is the black spots, round and irregular in form

Minor diseases

> **Anthracnose**: *Colletotrichum gloeosporioides* (Penz.) Penz. & Sacc.
> *Colletotrichum capsici* (Syd.) Butl. and Bisby

> **Dieback / Stem canker**: *Lasiodiplodia theobromae* Pat.

> **Powdery mildew**: *Pseudoidium jatrophae* (Braun & Cook 2012)

> **Fusarium wilt**: *Fusarium oxysporum* Schlecht. emend. Snyder & Hansen

> **Collar rot / root rot**: *Macrophomina phaseolina*

> ***Botrytis* rot**: *Botrytis ricini*

> *Jatropha mosaic India virus*

Chapter - 45

Diseases of Mulberry

S. No.	Disease	Pathogen
1.	Powdery mildew	*Phyllactinia corylea*
2.	Leaf rust	*Peridiospora mori*
3.	*Cercospora* leaf spot	*Cercospora moricola*
4.	*Pseudocercospora* leaf spot	*Pseudocercospora mori*
5.	Anthracnose	*Colletotrichum demantium*
6.	*Fusarium* Root Rot	*Fusarium oxysporum*
7.	Sooty mould	*Capnodium* sp.
8.	Bacterial leaf spot	*Pseudomonas syringae* pv. *mori*
9.	Mosaic	*Mulberry mosaic virus*

1. Powdery mildew

Symptoms

» White powdery patches appear on the lower surface of leaf which is gradually increased and cover whole leaf surface.

» Affected leaves turn yellowish to purplish brown and defoliate prematurely.

Diseases of Mulberry

Fungus: *Phyllactinia corylea* (Pers.) P. Karst.; *Phyllactinia guttata* (Wallr. ex. Fr). Lev.

Mycelium is superficial upon the lower surface of the leaf, septate with special intracellular haustoria and globed adhesive appresorium. The haustoria are finger like, irregular shaped. Conidiophores are ovuluriopsis type, straight, erect and hyaline. Conidia are subclavate, aseptate, solitary and hyaline.

Disease cycle

The pathogen is ectoparasitic, which obtains nutrients by haustoria. It reproduces by initially asexual, when matures follow sexual process. The conidia spread by wind current.

Favourable conditions

- » Poor fertility, aging and soil salinity.

- » Temperature 25°C and relative humidity 80%.

- » Cloudy rainy days favour the conidial formation and dry day weather favours its liberation.

Management

- » Grow resistant varieties like Mandania, Katania, China-White, Jodhpur, Punjab local, MR-1, MR-2, S 31, S 54, S 153, S 796, S 1096, Kanva-2, Calbresa, Cattaneo, Shrim-2, ACC123, ACC-125, ACC-151, ACC153, OPH-3, Almora local, Himachal local and 5-523.

- » Provide wider spacing 3' × 3' between rows and plants, keeping the single trunk of height 30 cm.

- » Spray Bordeaux mixture 1% or carbendazim 0.1% or morestan 0.025%.

2. Leaf rust

a) Peridiospora mori Barclay (Syn: Aecidium mori Barclay)

Symptoms

- » Several small pin head shaped, reddish brown, irregular pustules appear on the both surface of mature leaves and green woody portions of stem.

384   Diseases of Field Crops and their Management

» Severely infected leaves turn yellowish, margin of the leaves become dry and fall prematurely.

Fungus: *Peridiospora mori*: Aecia amphigenous, solitary or in groups, sometimes densely clustered, on leaves, buds, and branches, also on veins and petioles, often in elongated. Spermogonia, uredinia and telia are not reported.

b) Cerotelium fici (Butler) Arth.

Symptoms

» Raised rusty pustules appeared on the lower surface of the leaves.

» In severe, leaves are deformed and fall off.

Fungus: *Cerotelium fici*: It grows in the form of sub epidermal ostiolate uredinia. Urediospores are sessile, hyaline, globose, ellipsoidor obvoid.

Disease cycle

» The spores spread by wind current.

Favourable conditions

» Closer spacing and high humidity.

» Dry day condition and dewy night favours the disease development.

Management

» Grow resistant varieties like China Peking, Cattaneo and BC 259.

» Provide wider spacing 3' × 3' between rows and plants.

» Spray captafol 0.2% or chlorothalanil 0.2% or mancozeb 0.1% or propineb 0.2%.

3.   *Cercospora* leaf spot

Symptoms

» Minute, circular or irregular light brown spots appear on both sides of the leaves.

» The adjacent spots unite together to form a larger spot.

Diseases of Mulberry

» The necrotic tissues of such spots drop out and form the characteristics shot holes.

» Highly infected leaves become yellowing and defoliate prematurely.

Fungi: *Cercospora moricola* Cooke; *C. moricola, C. missouriensis* and *C. snelliana*

Mycelia produced in both internal and external part of host. Mycelia are filamentous and sepate. Conidiophores are dark, fasciculate, geniculate at the point of spore production. Conidia are simple, obclavate, hyaline and multiseptate.

Disease cycle

» The spores spread by rainsplashes and air.

Favourable conditions

» Closer planting.

» Temperature 24°C and relative humidity 90%.

Management

» Grow resistant varieties like Kalakuthai, Bidevalaya, Kanva-2, S-54, C-799, Shrim-2, Moulai, Mizusawa, Assamabola, Cattaneo, Miruso, ACC-106, ACC-109, ACC-122, ACC-114, ACC-115, ACC-116, ACC-117, ACC-119, ACC-121, ACC-123, ACC-125, ACC-128, ACC-152, ACC-153, ACC-201, OPH-1, OPH-3, MS-2, MS-3, MS-5, MS-7, MS-8, Paraguay, RFS-135, RFS-175, S-153, Almora Local and S-1096.

» Provide wider spacing.

» Spray carbendazim 0.025%.

4. *Pseudocercospora* **leaf spot**

Symptoms

» Small grey to brown colored hyphal mats on lower surface of the leaves, and deep brown colored spots develop on the upper surface.

» Lower most leaves exhibit yellowing and defoliation.

» The necrotic tissues of such spots drop out and form the characteristics shot holes.

386 Diseases of Field Crops and their Management

> Highly infected leaves become yellowing and defoliate prematurely.

Fungus: *Pseudocercospora mori* Hara

Mycelia are both internal and external part of host. Conidiophores are brown, simple or branched, septate, fasciculated. Conidia are pale brown, obcalvate and filamentous.

Disease cycle

> The pathogen survived in infected plant debris.

Favourable conditions

> Temperature 19-36°C.

Management

> Grow resistant varieties like C-763, C-1726, C-1635, Tr-4 and Tr-10.

> Spray carbendazim 0.1%.

5. Anthracnose

a) *Colletotrichum demantium* (Pers. Ex. Fr.) Grove (Syn: *C. morifolium* Hara)

Symptoms

> Initially faint brown lesions on leaf, which enlarge and transform into oval, angular or irregular shaped spots with black brown borders.

> It produces marginal scorching, drying, yellowing and fall prematurely.

Fungus: *Colletotrichum demantium*: Acervuli are black, circular, and subepidermal grown on lower surface of the infected leaves. Conidiophores are phailidic, conidia develop as simple. Conidia are hyaline, olive grey to salmon in mass and falcate.

b) *Colletotrichum gloeosporioides* (Penz.) Sacc. (Teleomorph: *Glomerella cingulata* Stone.)

Symptoms

> This is also known as black leaf spot disease.

Diseases of Mulberry 387

» Initially coral red to brown color spots appear on leaf blade, mostly towards the margin and along the veins.

» In severe, leaves are deformed and fall off.

Fungus: *Colletotrichum gloeosporioides*: Acervuli subepidermal at initial and become erumpent at late, round or irregular in shape. Setae are dark brown slightly swollen at the base and tapering towards the apex. Conidiophores are closely packed on the surface of the acervuli, faintly brown and cylindrical. Conidia are hyaline, pinkish in mass, aseptate and oblong to cylindrical.

b) *Colletotrichum lindemuthianum* (Sacc. & Magn.) Briosi and Cavara.

Symptoms

» Initially brown to reddish brown, round or oval and depressed eyespots on the leaf blade, spread towards the margin.

Fungus: *Colletotrichum lindemuthianum*: Acervuli are scattered, erumpent and elongate with black setae. Conidiophores are cylindrical, hyaline, faintly brown, simple and aseptate cylindrical. Conidia are hyaline, septate and oblong.

Disease cycle

» The spores spread by wind current.

» Initially form latent infection on immature leaves.

Favourable conditions

» Temperature 28-32°C and relative humidity 75-90%.

» Dry day condition and dewy night favours the disease development.

Management

» Grow resistant varieties like Kosen, S 146 and BC 259.

» Grow black leaf spot resistant varieties like China Peking, MR 2, ACC 151 and ACC 233.

» Spray *Bacillus amyloliquifaciens* commercial formulation.

» Spray carbendazim 0.025% or thiram 0.10% or mancozeb 0.2% or propineb 0.2%.

6. *Fusarium* Root Rot

Symptoms

» The leaves are suddenly withering and defoliation.

» The bark of the infected root is decayed and easily of peel off.

» The surface of root and cortex appears gummy and the plant dies.

Fungi

a. *Fusarium oxysporum* **Schlecht**: Sporodochia are convex, subverrucose, erumpent and confluent. Mycelium is white or peach with purple tinge, sparse, produces microconidia, macroconidia and resting chlamydospores in pairs or chains.

b. *Fusarium solani* (**Mart.**) **Sacc**: Sporodochia are globose, irregular and white. Mycelium is abuntant aerially, allantoid, produces microconidia, macroconidia and resting chlamydospores either terminal or intercalary.

Favourable conditions

» The disease prevalent in rainy season.

Management

» Diseased plants should be uprooted and burnt.

» The field should be flooded to destroy dormant mycelia and spores.

» Land reclaimed by placing chloropicrin or calcium cyanamide 2.5 kg / m².

» Soil disinfection by di-trapex or vorlex (80% chlorinated C3 hydrocarbon + 20% methyl isothiocyanide).

» Planting pits can be inoculated with bioagent (*Trichoderma harzianum* 1 kg + FYM 50 kg) at 500 g/ pit.

» Planting pits can be dusted with mancozeb 2%.

» Seedlings dip with carbendazim 0.2%.

» Provide wider spacing 3' × 3' between rows and plants.

» Spray captafol 0.2% or chlorothalanil 0.2% or mancozeb 0.1% or propineb 0.2%.

Diseases of Mulberry

7.   Sooty mould: *Capnodium* **sp.,** *Chaetothyrium* **sp.,** *Curvularia affinis*

Symptoms

> » Thick black coating developed on the upper surface of the leaves.

> » Associated with mulberry white fly (*Bemesia myricae*)

8.   Bacterial blight

Symptoms

a.   Shrunk leaf type or Halo blight

> » Green water soaked irregular spots appeared on the leaves, bacterial ooze out on the lower surface.

> » Later the spots become brown, blighted, irregular or slightly angular and surrounded by fade green or yellow halo.

> » In severe, leaves are wrinkled, distorted, shrunk backwards and eventually defoliate.

> » On the bark, appears as black, sunken and shuttle like lesions.

> » It is common during rainy season when there is high humidity and temperature.

b.   Blackish bacterial blight or Head rot or Plague.

> » It appears as semi-transparant spots on young shoots, later turn to blackish brown in color.

> » The xylem becomes brown and rotten.

Bacterium: *Pseudomonas syringae* **pv.** *mori*

> The bacterium is rod shaped, possess polar flagella, Gram-negative reaction.

Disease cycle

> » The pathogen overwinters on twigs and dormant buds.

Favourable conditions

> » Intermediate rain, temperature 24°C.

Management

390   Diseases of Field Crops and their Management

»   Grow resistant varieties like Kosen, S-36, S-41 and S-54.

»   Spray captafol 0.2% or mancozeb 0.2%.

9. Mosaic

a. *Mulberry mosaic virus*

»   Infected leaves shows typical mosaic mottling with slight curling and puckering.

Vector: *Aphis gossypii* Glover, *Myzus persicae* (Sulz.) and *Rhopalosiphum maidis* (Fitch) (Aphids).

b. *Fig mosaic virus*

»   Infected leaves shows faint chlorotic spots followed by systematic mottling.

Vector: *Aceria ficus* Cotte (Eriophyid mite).

Management

»   Grow resistant varieties like Ichinose, Kairyonezumegaishi of *M. alba* and Oshimaso and Kosen on *M. latifolia*.

»   Spray carbendazim 0.1%.

Minor diseases

»   **Fungal leaf blight:** *Alternaria alternata, Bipolaris tetramera, Fusarium pallidoroseum*

»   **Leaf blotch:** *Srirosporium mori*

»   **Leaf spot:** *Alternaria tenuissima* (Nees & T. Nees: Fr.) Wiltshire, *Phloeospora maculans* (Berenger) Allesch., *Coniothyrium albae Pandotra & A. Husain, Drechslera yamadi, Fusarium concolor* Reinking, *Fusarium solani* (Mart.) Sacc., *Phoma mororum* Saoo. Syll., *Phyllosticta morifolia* Passerini.

»   **Tar leaf spot:** *Myrothecium roridum* Tode ex Fr.

»   **Charcoal rot:** *Macrophomina phaseolina* (Tassi) Goid.

»   **Collar rot:** *Phoma mororum* Saoo. Syll, *Sclerotium rolfsii* Sacc.

»   **Die-back:** *Diaporthe nomurai* Hara, *Hendersonula toruloidea* Nattrass, *Phoma sorghina*.

Diseases of Mulberry 391

- » **Stem blight:** *Phloeospora ulmi* (Fr.) Wallr, *Phoma exigua* Desm.

- » **Stem canker:** *Lasiodiplodia theobromae* Pat.

- » **Stem rot:** *Claudopus nidulars, Ganoderma lucidum (Curtis)* P.Karst, *Sclerotinia libertana* (Kubota et al., 1966), *Tubercularia vulgaris* Tode.

- » **Twig blight:** *Coniothyrium foedans* Sacc, *Fusarium lateritium* Nees, *Fusarium moniliforme* (Sacc.) Nirenberg, *Fusarium roseum* Link, *Fusarium solani* (Mart.) Sacc., *Myxosporium diedickii* Syd.,, *Stigminia mori, Polyporus hispidus.*

- » **Root rot / wilt:** *Rhizoctonia bataticola* (Taub.) Butler (= *Macrophomina phaseolina*) (Tassi) Goidanich)

- » **Cotton root rot:** *Phymatotrichum omnivorum* (Duggar) Hennebert

- » ***Armillariella* root rot:** *Armillariella mellea* (Vahl) P. Karst.

- » **Voilet root rot:** *Helicobasidium mompa* Tanaka

- » **White root rot:** *Rosellinia necatrix* Prillieux

- » **Bacterial leaf spot:** *Xanthomonas campestris* pv. *mori* (Pommel) Dowson.

- » **Bacterial leaf scorch:** *Xylella fastidiosa* Wells et al.

- » **Bacterial wilt:** *Ralstonia solanacearum* (Smith 1896) Yabuuchi et al. 1996

- » **Bacterial rot I:** *Bacterium moricolum* (Yendo & Higushi)

- » **Bacterial rot II:** *Bacterium mori* (Boyer & Lambert)

- » **Soft rot:** *Erwinia carotovora* subsp. *carotovora*

- » **Yellow dwarf:** *Candidatus* Phytoplasma

- » *Mulberry latent virus*

- » *Mulberry ringspot virus*

- » *Tobacco necrotic virus*

- » *Mulberry yellow net vein virus*

- » *Potato spindle tuber viroid* (Syn: *Mulberry mosaic dwarf viroid*)

Chapter - 46

Fungal and Bacterial Spoilage of Grains and Storage Pathogens

Field fungi

These fungi invade the kernels or seeds before harvest, while the plants are growing in the field or after the grain are cut and swathed but before it is threshed. There are some exceptions to this, notably corn tored on the cob in cribs and exposed to the weather; under such circumstances it may be invaded by field fungi, or the field fungi already present in it may continue to grow. The predominant field fungi differ somewhat according to the crop, the region or geographic location, and the weather, but in wheat, rice, barley, and oats, grown in much of the world, the major field fungi that invade the kernels are species of *Alternaria, Cladosporium, Helminthosporium* and *Fusarium*.

» *Alternaria* is common in many grains and seeds, especially the cereals, but it is not restricted to the cereal seeds. It is, for example, a predominant fungus in peanuts.

» *Cladosporium* is common in cereal seeds that have been exposed to moist weather during harvest, especially grains harvested with the hulls on, such as barley, oats, and rice.

» *Helminthosporium* is common in many lots of cereal seeds especially if the weather just before harvest has been moist. It may cause discoloration of the seed, death of the germinating seed or young seedling or root rots and blights of the mature plant, but causes no loss in storage.

Fungal and bacterial spoilage of grains and storage pathogens

» *Fusarium* also is common in freshly harvested cereal seeds. Some strains or species of it cause "scab" in barley, wheat, and corn; scabby grain may be toxic to animals, including man, and so is undesirable for food or feed.

Major grain diseases associated with storage infections

1. Rice grain blast:

Anamorph: *Pyricularia oryzae* (Cavara) (Early: *Pyricularia grisea* Sacc)
Teleomorph: *Magnaporthe oryzae* (B.C. Couch).

2. Rice brown spot:

Anamorph: *Bipolaris oryzae* (Breda de Haan)
Syn: *Helminthosporium oryzae* (Breda de Haan)*,*
Drechslera oryzae (Dreshler)
Teleomorph: *Cochliobolus miyabeanus* (Ito and Kuribayashi).

3. Rice false smut:

Anamorph: *Ustilaginoidea virens* (Cooke, 1975) Tahahashi - Conidial stage
(Syn: *Claviceps oryzae – sativa* Hashioka) – Perithecial stage
Teleomorph: *Villosiclava virens.*

4. Rice kernel bunt:

Anamorph: *Tilletia barclayana* (Bref.) Sacc. & Syd. Syn: *Neovasia horrida* Bref.

5. Rice bacterial grain rot:

Pseudomonas glumae Kurita and Tabei

6. Rice bacterial blight:

Xanthomonas oryzae pv. *oryzae* (Ishiyama) Swings et al

7. Rice sheath brown rot:

Pseudomonas fuscovaginae Tanii et al.

8. **Rice foot rot:**

   Erwinia chrysanthemi Burkholder et al.

9. **Wheat loose smut:**

   Ustilago tritici (Pers.) Rostr.
   Syn: *Ustilago nuda* var. *tritici* G.W. Fisch. & C.G. Shaw

10. **Wheat bunts:**

    a. Rough spored bunt/ European bunt/ Hill bunt/ Stinking smut/ Common bunt
       - *Tilletia tritici* (Bjerk.) Syn: *Tilletia caries* (DC.) Tul. & C. Tul.

    b. *Smooth spored bunt*
       - *Tilletia laevis* J.G. Kühn Syn: *Tilletia foetida* (Wallr.)

    c. *Dwarf bunt*
       - *Tilletia controversa* J.G. Kühn.

11. **Wheat head blight:**

    Fusarium graminearum (Schwein.) Petch.

12. **Wheat glume blotch:**

    Phaeosphaeria nodorum (E. Müll.) Hedjar

13. **Wheat black point:**

    Bipolaris sorokiniana (Sacc.)

14. **Wheat black chaff:**

    Xanthomonas translucens pv. *undulosa* and *Pantoea agglomerans* (Ewing and Fife 1972) Gavini et al. 1989

15. **Wheat ear cockle:**

    Clavibacter tritici

Fungal and bacterial spoilage of grains and storage pathogens 395

16. Barley loose smut:

Ustilago tritici (Pers.) Rostr. *Ustilago nuda* (C.N. Jensen) Rostr., nom. nud.

17. Barley covered smut:

Ustilago hordei (Pers.) Lagerh

18. Barley black chaff and bacterial streak

Xanthomonas translucens pv. *translucens* (Jones, Johnson & Reddy) Vauterin, Hoste, Kersters & Swings

19. Barley bacterial kernel blight

Pseudomonas syringae pv. *syringae*

20. Barley basal glume rot

Pseudomonas syringae pv. *atrofaciens* (McCulloch) Young, Dye & Wilkie

21. Oat loose smut:

Ustilago avenae (Pers.) Rostr.

22. Oat covered smut:

Ustilago koelleri Wilbe.

23. Oat black chaff

Xanthomonas campestris pv. *translucens* (Jones, Johnson & Reddy) Vauterin, Hoste, Kersters & Swings

24. Oat bacterial blight (halo blight)

Pseudomonas coronafaciens pv. *coronafaciens*

25. Sorghum ergot:

Sphacelia sorghi McRae; *Claviceps sorghi* Kulkarni

Diseases of Field Crops and their Management

26. Sorghum grain smut:

Sphacelotheca sorghi (Link) Clint.

27. Sorghum loose smut:

Sphacelotheca cruenta (Kuhn) Potter.

28. Sorghum long smut:

Tolyposporium ehrenbergii (Kuhn) Pat

29. Sorghum head smut:

Sphacelotheca reliana (Kuhn) Clint.

30. Pearl millet ergot:

Claviceps fusiformis Loveless 1967; *Claviceps microcephala* (Waller) Tul.

31. Pearl millet smut:

Tolyposporium penicillariae Bref.

32. Maize smut:

a. Common smut - *Ustilago maydis* (DC.) Corda
   Syn: *Ustilago zeae* (Beckm.) Unger)

b. False smut - *Ustilaginoidea virens* (Cooke) Takah.

c. Head smut - *Sporisorium reilianum* J.G. Kühn Langdon & Full.
   Syn: *Sorosporium reilianum* (J. G. Kühn) McAlpine;
      Sphacelotheca reiliana (J. G. Kühn) G. P. Clinton.

33. Maize cobb rot:

Fusarium moniliforme Sheld.

34. Maize seed rot:

Bacillus subtilis

Fungal and bacterial spoilage of grains and storage pathogens

35. Maize bacterial stalk and top rot:

Erwinia carotovora subsp. *carotovora* (Jones) Bergey and *Erwinia chrysanthemi* pv. *zeae* (Sabet) Victoria, Arboleda & Muñoz

36. Maize bacterial stalk rot:

Enterobacter dissolvens (*Erwinia dissolvens* Burkholder)

37. Barnyard millet smut:

Ustilago panicifrumentacei Bref. and *Ustilago paradoxa* Syd. & But.

38. Ragi smut:

Ustilago eleusine Kulk; *Melanopsichium eleusinis* (Kulk) Mundk and Thirum.

39. Ragi ovary smut:

Ustilago eleusinis Kulkarni

40. Cotton boll rot:

Xanthomonas axonopodis *axanopodis* pv. *malvacearum*

41. Sunflower head rot:

Rhizopus arrhizus (Fisher), *Rhizopus stolonifer* (Ehrenb.) Vuill., *Rhizopus nigricans* Ehrenb. and *Rhizopus microspores* Tiegh.

Bacterial grain contaminants

»   *Salmonella, Escherichia coli* (Migula) Castellani & Chalmers and *Bacillus cereus Frankland & Frankland*

Storage fungi

The storage fungi comprise about a dozen species of *Aspergillus* (of which only about five are at all common), several species of *Penicillium, Sporendonema, Rhizopus, Mucor Alternaria, Aureobasidium, Cladosporium, Epicoccum, Fusarium, Helminthosporium* and *Claviceps*.

Major storage grain diseases

1. Discolored and spotted rice grains: the spikelets are dark brownish in color, upper surface showing black velvety appearance with brown spots. Some spots are spindle or eye shaped.

   Associated pathogens:
   > *Bipolaris oryzae*
   > *Pyricularia oryzae*
   > *Curvularia oryzae*
   > *Curvularia lunata*
   > *Cladosporium oryzae*
   > *Nigrospora oryzae*

2. Black point of wheat: The embryonic point of wheat seed become black and the brains are also shriveled.

   The main causal organism associated:
   > *Bipolaris sorokiniana*

   Other organisms may associate as saprophytes:
   > *Alternaria tenuis*
   > *Curvularia* spp.
   > *Fusarium* spp.

3. Discolored and shriveled light weight jute seeds: Due to attack of several pathogens, the seeds are discolored and shriveled.

   > *Macrophomina phaseolina*
   > *Botryodiplodia theohromae*
   > *Colletotrichum corchori*

4. Discolored and shriveled seeds of mustard: Due to attack of seed-borne pathogens, the seeds become discolored and shriveled.

   > *Alternaria brassicae* and *Alternaria brassicicola*
   > *Aspergillus* spp.
   > *Penicillium* spp.
   > *Phoma lingam*

Fungal and bacterial spoilage of grains and storage pathogens

5. Purple stained soybean seeds: Purple stain of soybean: The diseased seeds became purple in color in one side and other side is normal.

    Cercospora kikuchi

6. Discolored and shriveled seeds of mungbean:

    Causal organisms may associate:
Colletotrichum dematium	*Mungbean golden mosaic virus*
Cercospora canescens	*Mungbean mosaic virus*
Macrophomina phaseolina	*Mungbean mosaic virus*
Botryodiplodia theobromae	
Fusarium spp. *Phoma* spp.	

Storage yeasts: *Candida, Cryptococcus, Pichia, Sporobolomyces, Rhodotorula, Trichosporon*

Storage bacteria

Common phytogenic microorganisms include bacteria (e.g. *Pseudomonadaceae, Micrococcaceae, Lactobacillaceae and Bacillaceae*),

Mycotoxins produced by storage fungi

S. No.	Mycotoxin	Major Foods	Species	Health effects
1.	Aflatoxins	Maize, Groundnuts, Figs, Tree nuts (Aflatoxin M$_1$ (milk, milk products, meat)	*Aspergillus flavus* *Aspergillus parasiticus*	Hepatotoxic, carcinogenic, fatty liver
2.	Cyclopiazonic acid	Cheese, Maize, Groundnuts, Rodo millet	*Aspergillus flavus* *Penicillium aurantiogriseum*	Convulsions
3.	Deoxynivalenol	Cereals	*Fusarium graminearum*	Vomiting, food refusal
4.	T-2 toxin	Cereals	*Fusarium sporotrichioides*	Alimentary toxic aleukia
5.	Ergotamine	Rye	*Claviceps purpurea*	Neurotoxin
6.	Fumonisin	Maize	*Fusarium moniliforme*	Esophageal cancer

7.	Ochratoxin	Maize, Cereals, Peanuts, Coffee beans	*Penicilliumverrucosum, Aspergillus ochraceus*	Nephrotoxic, Hepatotoxic
8.	Patulin	Apple juice, damaged apples	*Penicillium expansum, P. patulum*	Edema, brain and possibly carcinogenic
9.	Penitrem	Walnuts	*Penicillum aurantiogriseum*	Tremors
10.	Sterigmatocystin	Cereals, Coffee beans, Cheese	*A. versicolor, A. flavus, Bipolaris, Chaetomium*	Hepatotoxic, carcinogenic
11.	Tenuazonic acid	Tomato paste	*Alternaria tenuis*	Convulsions,
12.	Citreoviridin	Rice	*Penicillium viridicatum*	Cardiac beri-beri
13.	Citrinin	Rice	*P. viridicatum, P. citrinum*	Nephrotoxin
14.	Zearolenone	Maize, Barley, Wheat	*Fusarium graminearum*	Oestrogenic

Management

» Adopt good agricultural practices.

» Maintaining optimum moisture percentage.

» Use Adathoda, Neem, Vitex, Pungam and Melia leaves in storage bins.

» Fumigation with formalin solution.

» Spray *Flavobacterium*

» Treating the bags and spraying on the floors with Carbendazim, Napam, Vapam, PMC, EMC and Captan.

References

- Agrios, G.N. 1998. Plant Pathology, Academic Press, New York.

- Alexopolus, C.J. and Mims. 1989. Introductory Mycology, Willey Eastern Ltd., New Delhi.

- Arjunan, G. Dinakaran, D. and Parthasarathy, S. 2019. Diseases of Horticultural Crops. Kaveri Publishers, Trichy.

- Biswas, S and Singh, N.P. 2005. Parasitic Diseases of Mulberry. Kalyani Publishers.

- Chattopadhyay, S.G. 1998. Principles and Procedure of Plant Protection – Oxford and IBH Publication, New Delhi.

- Dasgupta, M. K. 1988. Principles of Plant Pathology, Allied Publishers Pvt. Ltd., Bangalore

- Dasgupta, M.K. 1998. Principles of Plant Pathology, Allied Publishers Pvt. Ltd., Bangalore.

- Dickson, J.G. 1997. Diseases of Field Crops, Daya Publishing House, New Delhi.

- Dickson, J.G. 2008. Diseases of Field Crops. Biotech Books.

- Dube, H.C. 1992. A Text Book of Fungi, Bacteria and Viruses. Vikas publishing house Pvt. Ltd., New Delhi.

- Dubey, S.C. Rashmi Aggarwal, Patro T.S.S.K. and Pratibha Sharma. 2016. Diseases of Field Crops and their Management. Today and Tomorrows Printers and Publishers.

- Govindasamy, C.V. and Alagianalingam, M.N. 1981. Plant Pathology, Popular book Depot, Madras pp. 545.

- Gupta, V.K. and Paul. Y.S. 2002. Diseases of Field Crops. Indus Publishing Company.

- Kalita. M.K. 2014. Diseases of Field Crops and their Management. Kalyani Publishers.

- Kolte, S.J. 1985. Diseases of Annual Edible Oilseed Crops. Vol. III: Sunflower, Safflower and Nigerseed Diseases. CRC Press: Boca Raton, FL.

- Maramorach, K. 1998. Plant Diseases of Viral, Viroid, Mycoplasma and Uncertain Etiology, Oxford and IBM publications, New Delhi.

- Mehrota, R.S. 1980. Plant Pathology, Tata Mc Grow Mill Pub. Co. New Delhi. pp. 771.

- Mehrotra, R.S. 1990. An Introduction to Mycology, Willey Eastern Ltd., New Delhi.

- Narayanasamy, P. 1997. Plant Pathogens and Detection and Disease Diagnosis, CRC Publication, USA.

- Nene, Y.L. and Thapliyal, P.N. 1998. Fungicides in Plant Disease Control. Oxford and IBH Publishing Co. Ltd., New Delhi.

- Prakasam, V. Valluvaparidasan, V., Raguchander, T and Prabakar. K. 1997. Field Crops Diseases. AE Publication, Coimbatore. P 176

- Prakasam, V., Raguchander, T. and Prabakar, K. 1998. Plant Disease Management. A.E. Publication, Coimbatore.

- Rajamanickam, S. and Parthasarathy, S. 2017. Plant Pathology Refresher. Thannambikai Publications, Coimbatore.

- Rangaswami, G. 1998. Diseases of Crop plants in India. Prentice Hall of India Pvt.Ltd., New Delhi. pp. 504.

- Singh, R.S. 1993. Plant Diseases. Oxford & IBH Publication, New Delhi.

- Thind. T.S. 1998. Diseases of Field Crops and their Management. NATIC; 1st ed edition.

- Trivedi. 2011. Plant Health and their Management: Diseases of Pearl Millet and their Management. Agrobios (India).

- Vidyasekaran, P. 1993. Principles of Plant Pathology, CBS Publishers and Distributors, New Delhi.

- Wheeler, B.E.J. 1969. An Introduction to Plant Diseases – The English Language Book Society, London, United Kingdom.

- White, D.G. 1999. Compendium of Corn Diseases. Third Edition. St. Paul, MN: APS Press.